Contents

L

Use of guidance

THE APPROVED DOCUMENTS

The Building Regulations 1991 (SI 1991 No 2768), have been amended by the Building Regulations (Amendment) Regulations 1994 (SI 1994 No 1850). This document is one of a series that has been approved by the Secretary of State as practical guidance on meeting the requirements of Schedule 1 and regulation 7 of the Building Regulations.

At the back of this document is a list of those documents currently published by the Department of the Environment and the Welsh Office which have been approved for the purpose of the Building Regulations 1991.

The detailed provisions contained in the Approved Documents are intended to provide guidance for some of the more common building situations. In other circumstances, alternative ways of demonstrating compliance with the requirements may be appropriate.

Evidence supporting compliance

There is no obligation to adopt any particular solution contained in an Approved Document if you prefer to meet the relevant requirement in some other way. However, should a contravention of a requirement be alleged then, if you have followed the guidance in the relevant Approved Documents, that will be evidence tending to show that you have complied with the Regulations. If you have not followed the guidance, then that will be evidence tending to show that you have not complied. It will then be for you to demonstrate by other means that you have satisfied the requirement.

Other requirements

The guidance contained in an Approved Document relates only to the particular requirements of the Regulations which that document addresses. The building work will also have to comply with the requirements of any other relevant paragraphs in Schedule 1 to the Regulations. There are Approved Documents which give guidance on each of the other requirements in Schedule 1 and on regulation 7.

MATERIALS AND WORKMANSHIP

Any building work which is subject to requirements imposed by Schedule 1 of the Building Regulations should, in accordance with regulation 7, be carried out with proper materials and in a workmanlike manner.

You may show that you have complied with regulation 7 in a number of ways, for example, by the appropriate use of a product bearing an EC mark in accordance with the Construction Products Directive (89/106/EEC), or by following an appropriate technical specification (as defined in that Directive), a British Standard, a British Board of Agrément Certificate, or an alternative national technical specification of any member state of the European Community which, in use, is equivalent. You will find further guidance in the Approved Document supporting regulation 7 on materials and workmanship.

Technical specifications

Building Regulations are made for specific purposes: health and safety, energy conservation and the welfare and convenience of disabled people. Standards and technical approvals are relevant guidance to the extent that they relate to these considerations. However, they may also address other aspects of performance such as serviceability or aspects which although they relate to health and safety are not covered by the Regulations.

When an Approved Document makes reference to a named standard, the relevant version of the standard is the one listed at the end of the publication. However, if this version of the standard has been revised or updated by the issuing standards body, the new version may be used as a source of guidance provided it continues to address the relevant requirements of the Regulations.

The Secretary of State has agreed with the British Board of Agrément on the aspects of performance which it needs to assess in preparing its Certificates in order that the Board may demonstrate the compliance of a product or system which has an Agrément Certificate with the requirements of the Regulations. An Agrément Certificate issued by the Board under these arrangements will give assurance that the product or system to which the Certificate relates, if properly used in accordance with the terms of the Certificate, will meet the relevant requirements.

Similarly, the appropriate use of a product which complies with a European Technical Approval as defined in the Construction Products Directive will also meet the relevant requirements.

The Requirement L1

This Approved Document which takes effect on 1 July 1995 deals with the following Requirements which are contained in the Building Regulations 1991, as amended 1994.

Requirement	Limits on application
Conservation of fuel and power	
L1. Reasonable provision shall be made for the conservation of fuel and power in buildings by:	Requirements L1(a), (b), (c) and (d) apply only to –
(a) limiting the heat loss through the fabric of the building;	(a) dwellings;
(b) controlling the operation of the space heating and hot water systems;	(b) other buildings whose floor area exceeds 30 m².
(c) limiting the heat loss from hot water vessels and hot water service pipework;	
(d) limiting the heat loss from hot water pipes and hot air ducts used for space heating;	
(e) installing in buildings artificial lighting systems which are designed and constructed to use no more fuel and power than is reasonable in the circumstances and making reasonable provision for controlling such systems.	Requirement L1(e) applies only within buildings where more than 100 m² of floor area is to be provided with artificial lighting and does not apply within dwellings.

Other Changes to the Building Regulations 1991

Attention is particularly drawn to the following changes to the requirements in the Building Regulations 1991 which are brought into effect by the Building Regulations 1994 (Amendment) Regulations.

Requirements relating to material change of use.

Regulation 6(1)(a): the parts of Schedule 1 which are applicable in all cases.

The following line is added:

L1 (conservation of fuel and power).

New requirements for the provision of an energy rating

A new Regulation 14A is added as follows:

14A. - (1) This regulation applies where a new dwelling is created by building work or by a material change of use in connection with which building work is carried out.

(2) Where this regulation applies, the person carrying out the building work shall calculate the energy rating of the dwelling by means of a procedure approved by the Secretary of State, and shall give notice of that rating to the local authority.

(3) The notice shall be given –

(a) not later than the notice required by paragraph (4) of regulation 14, and

(b) when the occupation of a building or part of a building referred to in paragraph (5) of that regulation is occupation of a dwelling, not later than the notice referred to in that paragraph.

Note: This amendment applies to cases where a local authority is the building control body but the Department also proposes to make similar changes where the building control body is an Approved Inspector. The guidance in the Approved Document reflects both the amendments now made and those proposed.

General guidance

Performance

0.1   In the Secretary of State's view requirements L1(a), (b), (c) and (d) will be met by the provision of energy efficiency measures which:

a.   limit the heat loss through the roof, wall, floor, windows and doors, etc. and where appropriate permit the benefits of solar heat gains and more efficient heating systems to be taken into account;

b.   limit unnecessary ventilation heat loss by reducing air leakage around openings and through the building fabric;

c.   enable effective control of the operation of space heating and hot water systems by controlling temperatures and the duration of system operation;

d.   limit the heat loss from hot water vessels and heating and hot water pipes and ducts where such heat does not make an efficient contribution to the space heating.

0.2   In the Secretary of State's view requirement L1(e) will be met by the provision of lighting systems which:

a.   utilise energy-efficient artificial lighting sources;

b.   can be effectively controlled by manual and/or automatic switching to minimise their use and hence obtain the benefits of natural lighting.

Introduction to provisions

Small extensions to dwellings

0.3   Where an extension does not exceed 10 m² in floor area, reasonable provision may be considered to have been made if its construction is no less effective for the purposes of the conservation of fuel and power than the existing construction.

Buildings with low levels of heating

0.4   Because of the nature of their intended use some buildings may only require a low level of heating (or even no heating) and insulation of the building fabric will be unnecessary. As a general rule a low level of heating is where the output of the space heating system does not exceed 50 W/m² of floor area for industrial and storage buildings, or 25 W/m² of floor area for any other building which is not a dwelling.

0.5   Where the level of heating to be provided cannot be established (because, for instance, the use of the building when it is being constructed is not known) then insulation and sealing of the building fabric will be necessary.

Large complex buildings

0.6   In large complex buildings it may be sensible to consider the provisions for conservation of fuel and power separately for the different parts of the building in order to establish the measures appropriate to each part.

Technical risk

0.7   Guidance on avoiding technical risks (such as rain penetration, condensation, etc.) which might arise from the application of energy conservation measures is given in BRE Report BR 262 *Thermal insulation: avoiding risks* (although this is not an Approved Document). Guidance is also available in the NHBC publication *Thermal insulation and ventilation: Good Practice Guide* and Approved Document F also contains guidance on the provision of ventilation to reduce the risk of condensation.

Thermal conductivity and transmittance

0.8   The thermal conductivity (ie λ-value) of a material is a measure of the rate at which that material will pass heat and is expressed in units of Watts per metre per degree of temperature difference (W/mK). Thermal transmittance (ie the U-value) is a measure of how much heat will pass through one square metre of a structure when the air temperatures on either side differ by one degree. U-values are expressed in units of Watts per square metre per degree of temperature difference (W/m²K).

0.9   In the absence of certified manufacturers' information thermal conductivities (λ W/mK) and thermal transmittances (U W/m²K) may be taken from the tables in this Approved Document or alternatively in the case of U-values they may be calculated. If certified test results for particular materials and makes of products are available, however, they should be used in preference.

U-value reference tables

0.10   Table 2 on page 8 (repeated as Table 7 on page 17) contains indicative U-values for windows, doors and rooflights. Appendix A contains tables and examples of their use which provide a simple way to establish the amount of insulation needed to achieve a given U-value for some typical forms of construction. The values in the tables have been derived taking account of typical thermal bridging where appropriate.

Calculation of U-values

0.11   When calculating U-values the thermal bridging effects of, for instance, timber joists, structural and other framing, normal mortar bedding and window frames should generally be taken into account using the procedure given in Appendix B. Thermal bridging can be disregarded, however, where the difference in thermal resistance between the bridging material and the bridged material is less than 0.1 m²K/W. For example, normal mortar joints need not be taken into account in calculations for brickwork.

Basis for calculating areas

0.12   The dimensions for the areas of walls, roofs and floors should be measured between finished internal faces of the external elements of the building including any projecting bays. In the case of roofs they should be measured in the plane of the insulation. Floor areas should include non-usable space such as builders' ducts and stair wells.

Exposed and semi-exposed elements

0.13 In this document:

a. Exposed element means an element exposed to the outside air (including a suspended floor over a ventilated or unventilated void) or an element in contact with the ground.

b. Semi-exposed element means an element that separates a heated space from an unheated space which has exposed elements which do not meet the recommendations for the limitation of heat loss as described in Section 1 and Section 2. (See Diagram 1.)

Energy rating of dwellings

0.14 Dwellings provided as new construction or by way of material changes of use which include building work must be given energy ratings using the Government's Standard Assessment Procedure (SAP). An explanation of this procedure, appropriate reference data and a calculation worksheet are included in this Approved Document at Appendix G.

0.15 There is no obligation to achieve a particular SAP Energy Rating. However, higher levels of insulation are justified for new dwellings having SAP ratings of 60 or less, whereas a way of demonstrating compliance with the requirements could be to have a SAP rating of 80 to 85, dependent upon dwelling size.

0.16 The guidance in Section 1 has been formulated having regard to the Energy Rating to be achieved. When designing new dwellings it might be advisable to assess their SAP Energy Ratings at an early stage in design and to keep the estimates under review.

Calculations

0.17 SAP Energy Rating calculations undertaken by applicants should be submitted to the building control body.

0.18 Certification that SAP and other calculation procedures have been carried out correctly may be undertaken by a competent person and certified calculations may be accepted as such by building control bodies. The building control body will remain responsible for enforcement, however, and any question of competence should be settled prior to submission of certified calculations.

0.19 Individuals or bodies authorised by the Secretary of State as assessors for undertaking SAP calculations can be accepted by the building control body as competent persons for that purpose.

Certification of compliance

0.20 Where a person has been approved by the Secretary of State for the Environment as an "Approved Person" for the purpose of certifying compliance with Part L of the Building Regulations, work certified by that person will be accepted by the building control body.

Diagram 1 **Examples of semi-exposed elements**

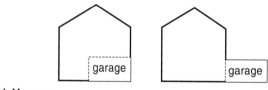

(a) Houses

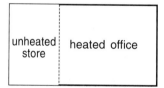

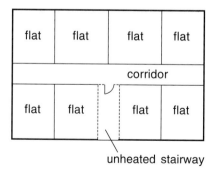

unheated stairway

(b) Flats: The exposed walls at the ends of the corridor are insulated and glazed so as to comply with the recommendations of Section 1. Therefore, walls that face into the corridor have no insulation requirements.

The exposed wall to the unheated stairway has a substantial glazed area and does not comply with the recommendations in Section 1. Therefore, the walls that face into the stairway are semi-exposed.

(c) Other buildings: The roof and the floor to the service duct do not comply with the recommendations of Section 2. Therefore, the walls of the duct are semi-exposed.

Key
Exposed element of construction **which is insulated to** recommended level is shown _____
Semi-exposed element is shown ------------------

Section 1

DWELLINGS

Insulation of the building fabric

Alternative methods of showing compliance

1.1 Three methods are shown for demonstrating how heat loss through the building fabric should be limited:

a. An **Elemental** method.

b. A **Target U-value** method.

c. An **Energy Rating** method.

Elemental method

Standard U-values for construction elements

1.2 The requirement will be met if the thermal performances of the construction elements conform with Table 1. The U-values in column (a) should be used if the SAP rating will be 60 or less as determined in accordance with the procedure in Appendix G. Where the SAP rating will be more than 60 the U-values in column (b) can be used. The standard U-values for windows, doors and rooflights may be modified in accordance with paragraph 1.10.

1.3 One way of achieving the U-values in Table 1 is by providing insulation of an appropriate thickness estimated from the tables in Appendix A. Examples of the SAP ratings of several typical dwelling designs are included in Appendix G to assist the initial assessment of new designs. An alternative procedure for demonstrating satisfactory provision for floors is given in Appendix C.

Table 1 **Standard U-values** (W/m²K) **for dwellings**

Element	For SAP Energy Ratings of:	
	60 or less (a)	over 60 (b)
Roofs [1]	0.2	0.25 [2]
Exposed walls	0.45	0.45
Exposed floors and ground floors	0.35	0.45
Semi-exposed walls and floors	0.6	0.6
Windows, doors and rooflights	3.0	3.3

Notes

1. Any part of a roof having a pitch of 70° or more may have the same U-value as a wall.

2. For a flat roof or the sloping parts of a room-in-the-roof construction it will be acceptable if a U-value of 0.35 W/m²K is achieved.

Windows, doors and rooflights

BASIC ALLOWANCE

1.4 The requirement will be met if the average U-value of the windows, doors and rooflights does not exceed the appropriate figure in Table 1 and the area of the windows, doors and rooflights together does not exceed 22.5% of total *floor area*. The U-values can be achieved by windows having sealed double-glazed units or by other systems (such as secondary glazing) which incorporate two or more panes of glass or other glazing material with space between.

1.5 The average U-value of windows, doors and rooflights in extensions to existing dwellings should not exceed 3.3 W/m²K. To establish the appropriate area of windows, doors and rooflights for

Table 2 **Indicative U-values** (W/m²K) **for windows, doors and rooflights**

Item	Type of frame							
	Wood		**Metal**		**Thermal break**		**PVC-U**	
	6	12	6	12	6	12	6	12
Air gap in sealed unit (mm)	6	12	6	12	6	12	6	12
Window, double-glazed	3.3	3.0	4.2	3.8	3.6	3.3	3.3	3.0
Window, double-glazed, low-E	2.9	2.4	3.7	3.2	3.1	2.6	2.9	2.4
Window, double-glazed, Argon fill	3.1	2.9	4.0	3.7	3.4	3.2	3.1	2.9
Window, double-glazed, low-E, Argon fill	2.6	2.2	3.4	2.9	2.8	2.4	2.6	2.2
Window, triple-glazed	2.6	2.4	3.4	3.2	2.9	2.6	2.6	2.4
Door, half double-glazed	3.1	3.0	3.6	3.4	3.3	3.2	3.1	3.0
Door, fully double-glazed	3.3	3.0	4.2	3.8	3.6	3.3	3.3	3.0
Rooflights, double-glazed at less than 70° from horizontal	3.6	3.4	4.6	4.4	4.0	3.8	3.6	3.4
Windows and doors, single-glazed	4.7		5.8		5.3		4.7	
Door, solid timber panel or similar	3.0		—		—		—	
Door, half single-glazed, half timber or similar	3.7		—		—		—	

extensions, however, the basic allowance in paragraph 1.4 can be applied to either:

a. the *floor area* of the extension itself; or

b. the *floor area* of the existing dwelling and extension together.

1.6 Door designs can include various panel arrangements, but the indicative U-values given in Table 2 will generally be acceptable. Single-glazed panels would be acceptable in external doors provided they do not increase the average U-value for windows, doors and rooflights beyond the limit dependent upon the area of openings as obtained from Table 3.

1.7 Windows and doors with single-glazed panels protected by unheated, enclosed, draught-proof porches or conservatories may be assumed for the purposes of Building Regulations to have a U-value of 3.3 W/m²K.

Table 3 **Permitted variation in the area of windows and doors for dwellings**

Average U-value (W/m²K)	Maximum permitted area of windows and doors as a percentage of floor area for SAP Energy Ratings of:	
	60 or less	over 60
2.0	37.0%	41.5%
2.1	35.0%	39.0%
2.2	33.0%	36.5%
2.3	31.0%	34.5%
2.4	29.5%	33.0%
2.5	28.0%	31.5%
2.6	26.5%	30.0%
2.7	25.5%	28.5%
2.8	24.5%	27.5%
2.9	23.5%	26.0%
3.0	**22.5%**	25.0%
3.1	21.5%	24.0%
3.2	21.0%	23.5%
3.3	20.0%	**22.5%**
3.4	19.5%	21.5%
3.5	19.0%	21.0%
3.6	18.0%	20.5%
3.7	17.5%	19.5%
3.8	17.0%	19.0%
3.9	16.5%	18.5%
4.0	16.0%	18.0%
4.1	15.5%	17.5%
4.2	15.5%	17.0%

Note: The data in this table is derived assuming a constant heat loss through the elevations amounting to the loss when the basic allowance for openings of 22.5% of floor area is provided and the standard U-values given in Table 1 apply. It is also assumed for the purposes of this table that there are no rooflights.

1.8 Care should be taken in the selection and installation of appropriate sealed double-glazed windows in order to avoid the risk of condensation forming between the panes. Guidance on avoiding this problem is given in BRE Report BR 262 *Thermal insulation: avoiding risks.*

MODIFICATION TO THE BASIC ALLOWANCE
1.9 The percentage area allowance in paragraph 1.4 is based on average U-values as given in Table 1. The average U-value will depend upon the individual U-values of the components proposed and their proportion of the total area of openings as illustrated in Appendix E. In the absence of certified manufacturers' data the indicative U-values for components given in Table 2 can be used. If certified manufacturers' data is available, however, it should be used in preference.

1.10 Areas of windows, doors and rooflights larger than that given in paragraph 1.4 may be adopted provided there is a compensating improvement in the average U-value. Table 3 indicates the variation in the area of openings which can be achieved within this constraint.

Summary of provisions in the Elemental method
1.11 Diagram 2 summarises the fabric insulation standards and allowances for windows, doors and rooflights given in the Elemental method. Examples of the procedures used in this method are given in Appendices A, C and E.

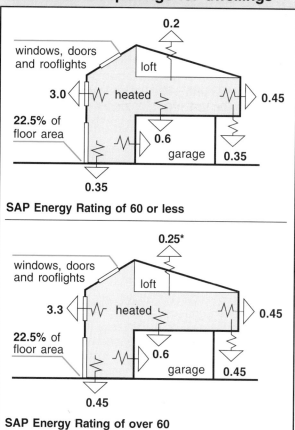

Diagram 2 **Standard U-values** (W/m²K) **and areas of openings for dwellings**

windows, doors and rooflights — 0.2 — loft — 3.0 heated — 0.45 — 22.5% of floor area — 0.6 — garage — 0.35 — 0.35

SAP Energy Rating of 60 or less

windows, doors and rooflights — 0.25* — loft — 3.3 heated — 0.45 — 22.5% of floor area — 0.6 — garage — 0.45 — 0.45

SAP Energy Rating of over 60
* 0.35 where there is no loft

Target U-value method

1.12 The requirement will be met if the calculated average U-values do not exceed the following targets:

a. For dwellings with SAP Energy Ratings of 60 or less:

$$\text{Target U-value} = \frac{\text{total floor area x 0.57}}{\text{total area of exposed elements}} + 0.36$$

b. For dwellings with SAP Energy Ratings of more than 60:

$$\text{Target U-value} = \frac{\text{total floor area x 0.64}}{\text{total area of exposed elements}} + 0.4$$

1.13 The total area of exposed elements should be calculated in accordance with paragraphs 0.12 and 0.13 and will comprise:

a. the area of fabric including windows, doors and rooflights which is exposed externally to outside air, plus

b. the area of the ground floor.

1.14 Semi-exposed elements should be insulated at least to the standard given in Table 1 and omitted from the calculation of the average U-value.

1.15 Example calculations for Target U-values and average U-values are given in Appendix F.

Optional method for accounting for solar gains
1.16 The Target U-value equations are based on a calculation assuming equal distribution of glazing on north and south elevations. Where the area of glazing on the south elevation exceeds that on the north, the total window area included in the calculation can be reduced as a way of taking account of the benefits of solar gains. It can be taken as the actual window area less 40% of the difference in area of glazing facing south ±30° and glazing facing north ±30°. Appendix F includes an example of the use of this procedure.

Optional method for accounting for higher efficiency heating systems
1.17 The Target U-value is based on a calculation assuming a gas or oil fired hot water central heating system with a seasonal efficiency of at least 72%. If a dwelling is to be heated in a more efficient manner (having regard for both heating system efficiency and primary energy consumption) a proportion of the benefits which will be obtained could be taken into account by increasing the Target U-value by up to 10%.

1.18 If the boiler in the above system were to be replaced by a condensing boiler, for example, this would increase seasonal efficiency to 85% and the Target U-value could be increased by the 10% allowed. As another example, a high efficiency electrical heating system comprising a heat pump with a seasonal coefficient of performance of 2.5 (taking account of any heating distribution losses) and a mechanical ventilation system with heat recovery would also enable the Target U-value to be increased by 10%. Other heating systems which provide an intermediate improvement in seasonal efficiency between 72% and 85% would justify a smaller increase in the Target U-value pro rata. Appendix F includes an example calculation which illustrates this procedure.

Energy Rating method

1.19 This is a calculation for dwellings using the Government's Standard Assessment Procedure as given in Appendix G which allows the use of any valid energy conservation measures. The procedure takes account of ventilation rate, fabric losses, water heating requirements, internal heat gains and solar gains. The requirement will be met if the SAP Energy Rating for the dwelling (or each dwelling in a block of flats or converted building) is not less than the appropriate figure shown in Table 4.

Table 4 SAP Energy Ratings to demonstrate compliance

Dwelling floor area (m²)	SAP Energy Rating
80 or less	80
more than 80 up to 90	81
more than 90 up to 100	82
more than 100 up to 110	83
more than 110 up to 120	84
more than 120	85

Using the calculation procedures

Limiting U-values
1.20 When using the calculation procedures in the **Target U-value** and **Energy Rating** methods it may be possible to achieve satisfactory solutions where the U-values of some elements are worse than those set out in Table 1. However, as a general rule the U-values of exposed walls and exposed floors should not be worse than 0.7 W/m²K and the U-value of roofs should not be worse than 0.35 W/m²K. U-values should be calculated by the method described in paragraph 0.11.

Limiting thermal bridging
1.21 When using the **Energy Rating** method the designs for lintels, jambs and sills should perform no worse than recommended in paragraphs 1.23 and 1.24.

Thermal bridging around openings

1.22 Provision should be made to limit the thermal bridging which occurs around windows, doors and other wall openings. This is necessary in order to avoid excessive additional heat losses and the possibility of local condensation problems.

1.23 Lintel, jamb and sill designs similar to those shown in Diagram 3 would be satisfactory and heat losses due to thermal bridging can be ignored if they are adopted.

Diagram 3 **Reducing thermal bridging around openings**

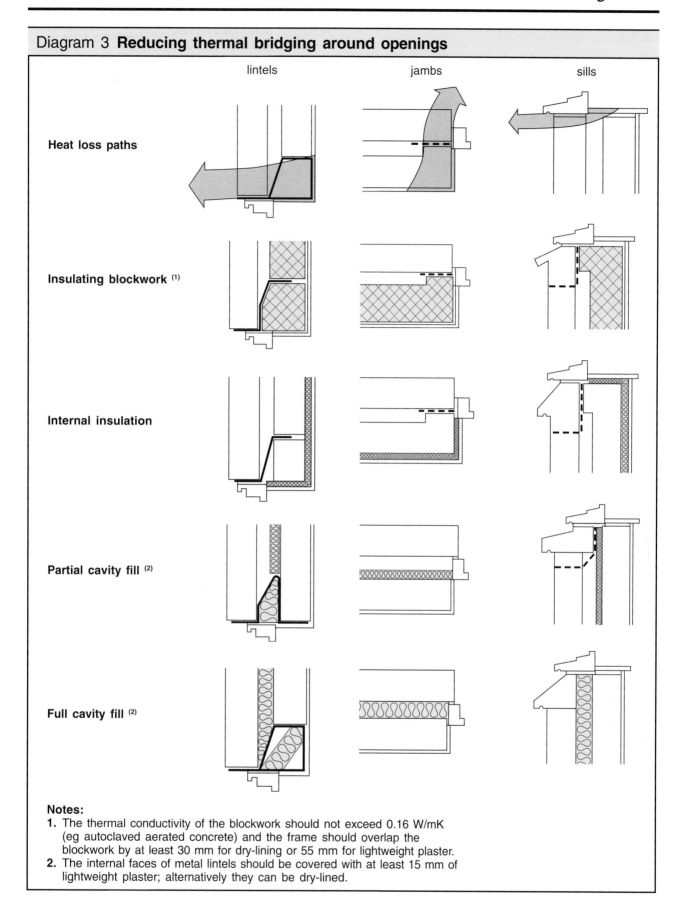

	lintels	jambs	sills
Heat loss paths			
Insulating blockwork [1]			
Internal insulation			
Partial cavity fill [2]			
Full cavity fill [2]			

Notes:

1. The thermal conductivity of the blockwork should not exceed 0.16 W/mK (eg autoclaved aerated concrete) and the frame should overlap the blockwork by at least 30 mm for dry-lining or 55 mm for lightweight plaster.

2. The internal faces of metal lintels should be covered with at least 15 mm of lightweight plaster; alternatively they can be dry-lined.

Alternative method

1.24 An alternative way of demonstrating compliance would be to show by calculation that the edge details around openings will give a satisfactory performance. Appendix D gives a procedure for this.

Limiting infiltration

1.25   Space heating demand is significantly affected by infiltration of cold outside air through leakage paths in the building envelope. It is therefore desirable to limit leakage by reducing unintentional air paths as far as is practicable. One way of satisfying the requirements would be to provide the following measures:

a.   sealing the gaps between dry-lining and masonry walls at the edges of openings such as windows and doors, and at the junctions with walls, floors and ceilings (eg by continuous bands of fixing plaster);

b.   sealing vapour control membranes in timber-frame constructions;

c.   fitting draught-stripping in the frames of openable elements of windows, doors and rooflights;

d.   sealing around loft hatches;

e.   ensuring boxing for concealed services is sealed at floor and ceiling levels, and sealing piped services where they penetrate or project into hollow constructions or voids.

1.26   Diagram 4 illustrates sealing measures which would satisfy the requirement. Further guidance is given in BRE Report BR 262 *Thermal insulation: avoiding risks* and in the NHBC publication *Guide to thermal insulation and ventilation*.

Space heating system controls

1.27   The guidance covers provisions which are appropriate for the more common varieties of heating system, excluding space heating provided by individual solid fuel, gas and electric fires or room heaters which have integral controls.

1.28   The requirement will be met by the appropriate provision of:

a.   zone controls; and

b.   timing controls; and

c.   boiler control interlocks.

Zone controls

1.29   These would be appropriate for hot water central heating systems, fan controlled electric storage heaters and electric panel heaters to control the temperatures independently in those areas (such as separate sleeping and living areas) that require different temperatures. The control devices could be room thermostats and/or thermostatic radiator valves or any other suitable temperature sensing devices.

1.30   In most dwellings two zones would be appropriate. However, in single storey open plan flats and bedsitters, for example, the heating system could be controlled in a single zone.

1.31   Zone controls would not be appropriate for ducted warm air systems and flap controlled electric storage heaters but these systems should have thermostats.

Diagram 4 **Limiting infiltration in dwellings**

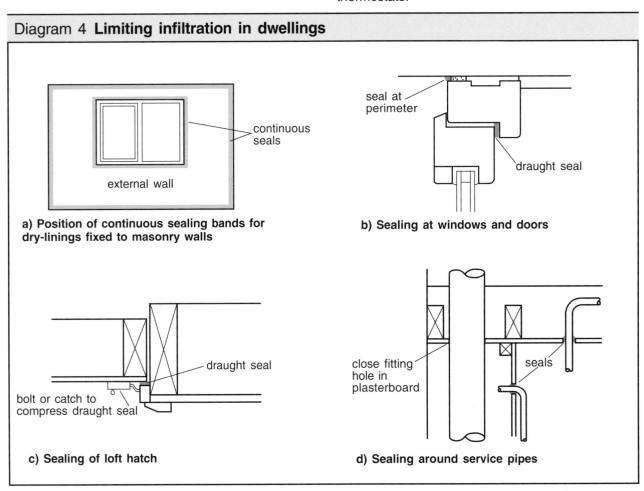

a) Position of continuous sealing bands for dry-linings fixed to masonry walls

continuous seals

external wall

b) Sealing at windows and doors

seal at perimeter

draught seal

c) Sealing of loft hatch

draught seal

bolt or catch to compress draught seal

d) Sealing around service pipes

close fitting hole in plasterboard

seals

Timing controls

1.32 Timing devices should be provided to control the periods when the heating systems operate. This provision should be made for gas fired and oil fired systems and for systems with solid fuel fired boilers where forced-draught fans operate when heat is required. Timing systems would be inappropriate for systems with solid fuel boilers which operate only by natural draught.

Boiler control interlocks

1.33 Gas and oil fired hot water central heating system controls should switch the boiler off when no heat is required whether control is by room thermostats or by thermostatic radiator valves:

a. Systems controlled by thermostats should fire only when a space heating or cylinder thermostat is calling for heat.

b. Systems controlled by thermostatic radiator valves should be fitted with flow control or other devices to prevent unnecessary boiler cycling.

Hot water storage system controls

1.34 For a system other than one heated by a solid fuel fired boiler the requirement will be met if:

a. the heat exchanger in the storage vessel has sufficient heating capacity for effective control: a way of satisfying this requirement would be to provide vessels complying with BS 1566 or BS 3198 or equivalent, and in particular with the requirements for the surface areas and pipe diameters of heat exchangers given in these Standards;

b. a **thermostat** is provided which shuts off the supply of heat when the storage temperature is reached, and which in the case of a hot water central heating system is interconnected with the room thermostat(s) to switch off the boiler when no heat is required; and

c. a **timer** is provided either as part of the central heating system or as a local device which enables the supply of heat to be shut off for the periods when water heating is not required.

1.35 For systems with solid fuel fired boilers where the cylinder is not providing the slumber load the requirement will be met by the provision of a **thermostatically controlled valve**.

Alternative approaches

1.36 The requirements for space heating and hot water storage system controls may be met by adopting the relevant recommendations in the following standards provided they include zoning, timing and anti-cycling control features similar to the above:

a. BS 5449: 1990 *Specification for forced circulation hot water central heating systems for domestic premises*;

b. BS 5864: 1989 *Specification for installation in domestic premises of gas-fired ducted air-heaters of rated output not exceeding 60 kW*;

c. other authoritative design specifications recognised by the heating fuel supply company.

Insulation of vessels, pipes and ducts

Insulation of hot water vessels

1.37 For a 120 litre vessel (450 mm dia by 900 mm high) the requirement will be met by providing factory applied insulation that restricts standing heat losses to 1 W/litre or less in tests using the method in BS 1566 Part 1, Appendix B.4 or equivalent. For other vessel sizes the requirement will be met by providing the same insulation in terms of both material and thickness as required for a 120 litre vessel or equivalent. A way of satisfying the requirement would be to provide vessels with a 35 mm thick, factory-applied coating of PU-foam having zero ozone depletion potential and a minimum density of 30 kg/m .

1.38 For unvented hot water systems additional insulation should be provided to control the heat losses through the safety fittings and pipework but without impeding safe operation and visibility of warning discharges. See Approved Document G3.

Insulation of pipes and ducts

1.39 Unless the heat loss from a pipe or duct contributes to the useful heat requirement of a room or space the pipe or duct should be insulated. The requirement would be satisfied if:

a. for pipes the insulation material has a thermal conductivity not exceeding 0.045 W/mK and a thickness equal to the outside diameter of the pipe up to a maximum of 40 mm, see Diagram 5; OR,

b. for pipes, and in the case of warm air ducts, insulation is provided in accordance with the recommendations of BS 5422: 1990 *Methods for specifying thermal insulation materials on pipes, ductwork and equipment.*

Diagram 5 **Insulation of heating and hot water pipes and warm air ducts**

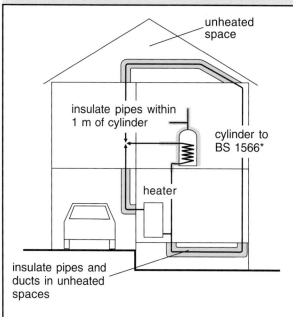

unheated space

insulate pipes within 1 m of cylinder

cylinder to BS 1566*

heater

insulate pipes and ducts in unheated spaces

* Normally a 120 litre cylinder would have a coil with about five turns

---

Conservation of fuel and power

1.40 The hot pipes connected to hot water storage vessels including the vent pipe, and the primary flow and return to the heat exchanger, where fitted, should be insulated for at least one metre from their points of connection or they should be insulated up to the point where they become concealed. The insulation should comprise 15 mm of a material having a thermal conductivity of 0.045 W/mK or other material applied in a thickness giving equivalent performance.

1.41 It should be noted that central heating and hot water pipework in unheated areas may need increased insulation thicknesses for the purpose of protection against freezing. Guidance on suitable protection measures is given in BRE Report BR 262 *Thermal insulation: avoiding risks.*

Conservatories

1.42 A conservatory has not less than three-quarters of the area of its roof and not less than one-half of the area of its external walls made of translucent material.

1.43 When a conservatory is attached to and built as part of a new dwelling:

a. where there is no separation between the conservatory and the dwelling the conservatory should be treated as an integral part of the dwelling;

b. where there is separation between the conservatory and the dwelling energy savings can be achieved if the conservatory is not heated. If fixed heating installations are proposed, however, they should have their own separate temperature and on/off controls.

1.44 For the purposes of satisfying the requirements for the conservation of fuel and power, separation between a dwelling and a conservatory means:

a. separating walls and floors insulated to at least the same degree as semi-exposed walls and floors;

b. separating windows and doors with the same U-value and draught-stripping provisions as the exposed windows and doors elsewhere in the dwelling.

1.45 Attention is drawn to the safety requirements of Part N of the Building Regulations regarding conservatory glazing.

Material alterations

1.46 Material alterations are defined in Regulation 3(2) as follows:

An alteration is material for the purposes of these Regulations if the work, or any part of it, would at any stage result –

(a) in a building or controlled service or fitting not complying with a "relevant requirement" where previously it did; or

(b) in a building or controlled service or fitting which before the work commenced did not comply with a "relevant requirement", being more unsatisfactory in relation to such a requirement.

1.47 "Relevant requirement" is defined in Regulation 3(3) as follows:

In paragraph 3(2) "relevant requirement" means any of the following applicable requirements of Schedule 1, namely:

Part A (structure)

paragraph B1 (means of escape)

paragraph B3 (internal fire spread – structure)

paragraph B4 (external fire spread)

paragraph B5 (access and facilities for the fire service)

Part M (access and facilities for disabled people).

1.48 When undertaking a material alteration the requirement may be satisfied in the following ways, although the extent of provision will depend upon the circumstances in each case:

a. roof insulation: when substantially replacing a roof structure – providing insulation to achieve the U-value for new dwellings;

b. floor insulation: where the structures of ground floors are to be substantially replaced – providing insulation in heated rooms to the standard for new buildings;

c. wall insulation: when substantially replacing complete external walls – providing a reasonable thickness of insulation and incorporating sealing measures as indicated in paragraph 1.25;

d. when carrying out building work on **space heating and hot water** systems – providing controls and insulation in accordance with paragraphs 1.28 to 1.41 as if they are new installations.

Material changes of use

1.49 Material changes of use are defined in Regulation 5 as follows:

For the purposes of ... these Regulations, there is a material change of use where there is a change in the purposes for which or the circumstances in which a building is used, so that after that change –

(a) the building is used as a dwelling, where previously it was not;

(b) the building contains a flat, where previously it did not; ...

1.50 When undertaking a material change of use the requirement may be satisfied in the following ways, although the extent of the provision will depend upon the circumstances in each case:

a. upgrading the insulation in accessible lofts: for example, additional insulation should generally be provided to achieve a U-value not exceeding 0.35 W/m²K where the existing insulation provides a U-value worse than 0.45 W/m²K;

b. **roof insulation**: when substantially replacing a roof structure – providing insulation to achieve the U-value for new dwellings;

c. **floor insulation**: where the structures of ground floors are to be substantially replaced – providing insulation to the standard for new buildings in heated rooms;

d. **wall insulation**: when substantially replacing complete exposed walls – providing a reasonable thickness of insulation and incorporating sealing measures as indicated in paragraph 1.25;

e. **wall insulation**: where internal surfaces are to be renovated over a substantial area – upgrading the insulation of exposed and semi-exposed walls. This could be achieved by providing a reasonable thickness of insulating dry-lining sealed in accordance with the guidance in paragraph 1.25;

f. **windows**: where windows are to be replaced – providing new draught-stripped windows with an average U-value not exceeding 3.3 W/m^{2}K. This could be inappropriate in conservation work and other situations where the existing window design needs to be retained;

g. when carrying out building work on **space heating and hot water** systems – providing controls and insulation in accordance with paragraphs 1.28 to 1.41 of this Approved Document as if they are new installations.

Section 2

BUILDINGS OTHER THAN DWELLINGS

Insulation of the building fabric

Alternative methods of showing compliance
2.1 Three methods are given for demonstrating how heat loss through the building fabric should be limited:

a. An **Elemental** method.

b. A **Calculation** method.

c. An **Energy Use** method.

Elemental method

Standard U-values for construction elements
2.2 The requirement will be met if the thermal performances of the construction elements conform with Table 5. One way of achieving the U-values in Table 5 is by providing insulation of a thickness estimated from the tables in Appendix A as illustrated in the examples. An alternative procedure for demonstrating satisfactory provision for floors is given in Appendix C.

Table 5 Standard U-values (W/m²K) for buildings other than dwellings

Element	U-value
Roofs [1]	0.25 [2]
Exposed walls	0.45
Exposed floors and ground floors	0.45
Semi-exposed walls and floors	0.6
Windows, personnel doors and rooflights	3.3
Vehicle access and similar large doors	0.7

Notes
1. Any part of a roof having a pitch of 70° or more may have the same U-value as a wall.
2. For a flat roof or insulated sloping roof with no loft space it will be acceptable if a U-value of 0.35 W/m²K is achieved for residential buildings or 0.45 W/m²K for other buildings.

Windows, doors and rooflights
BASIC ALLOWANCE
2.3 The requirement will be met if the average U-value of the windows, personnel doors and rooflights does not exceed the figure in Table 5 and the areas of the windows, personnel doors and rooflights do not exceed the percentages given in Table 6. Display windows, shop entrance doors and similar glazing may be excluded in calculations, however, for the purposes of the conservation of fuel and power.

2.4 The U-value of 3.3 W/m²K can be achieved by windows having sealed double-glazed units or by other systems (such as secondary glazing) which incorporate two or more panes of glass or other glazing material with space between.

2.5 The average U-value of windows, personnel doors and rooflights in extensions to existing buildings should not exceed 3.3 W/m²K. However, to establish the appropriate area of windows, personnel doors and rooflights for extensions the relevant basic allowance in Table 6 can be applied to:

a. the *wall area* of the extension itself; or

b. the *wall area* of the existing building and extension together.

2.6 Personnel door designs can include various panel arrangements but the indicative U-values given in Table 7 will generally be acceptable. Single-glazed panels would be acceptable in external personnel doors provided they do not increase the average U-value for windows, personnel doors and rooflights beyond the limit dependent upon the area of openings as obtained from Table 8.

2.7 Windows and personnel doors with single-glazed panels which are protected by unheated, enclosed, draught-proof lobbies may be assumed for the purposes of Building Regulations to have a U-value of 3.3 W/m²K.

2.8 Care should be taken in the selection and installation of appropriate sealed double-glazed windows in order to avoid the risk of condensation forming between the panes. Guidance on avoiding this problem is given in BRE Report BR 262 *Thermal insulation: avoiding risks.*

Table 6 Basic allowance for windows, doors and rooflights for buildings other than dwellings

Building type	Windows and doors [1]	Rooflights
Residential buildings[2]	30%	
Places of assembly, offices and shops	40% [3]	20% of roof area
Industrial and storage buildings	15%	
Vehicle access doors (all building types)	As required	

Notes
1. Percentage of exposed wall area.
2. Residential buildings (other than dwellings) means buildings in which people temporarily or permanently reside: for example, institutions, hotels and boarding houses.
3. See paragraph 2.3 regarding exclusions.

Table 7 Indicative U-values (W/m²K) for windows, doors and rooflights

Item	Type of frame							
	Wood		Metal		Thermal break		PVC-U	
Air gap in sealed unit (mm)	6	12	6	12	6	12	6	12
Window, double-glazed	3.3	3.0	4.2	3.8	3.6	3.3	3.3	3.0
Window, double-glazed, low-E	2.9	2.4	3.7	3.2	3.1	2.6	2.9	2.4
Window, double-glazed, Argon fill	3.1	2.9	4.0	3.7	3.4	3.2	3.1	2.9
Window, double-glazed, low-E, Argon fill	2.6	2.2	3.4	2.9	2.8	2.4	2.6	2.2
Window, triple-glazed	2.6	2.4	3.4	3.2	2.9	2.6	2.6	2.4
Door, half-double-glazed	3.1	3.0	3.6	3.4	3.3	3.2	3.1	3.0
Door, fully double-glazed	3.3	3.0	4.2	3.8	3.6	3.3	3.3	3.0
Rooflights, double-glazed at less than 70° from horizontal	3.6	3.4	4.6	4.4	4.0	3.8	3.6	3.4
Windows and doors, single-glazed	4.7		5.8		5.3		4.7	
Door, solid timber panel or similar	3.0		—		—		—	
Door, half-single-glazed, half timber panel or similar	3.7		—		—		—	

MODIFICATION TO THE BASIC ALLOWANCE

2.9 The percentage area allowances given in Table 6 are based on an average U-value of 3.3 W/m²K. The average U-value will depend upon the individual U-values of the components proposed and their proportion of the total area of openings as illustrated in Appendix E. In the absence of certified manufacturers' data the indicative U-values for components given in Table 7 can be used. If certified manufacturers' data is available, however, it should be used in preference.

2.10 Areas of windows, personnel doors and rooflights larger than those in Table 6 may be adopted provided there is a compensating improvement in the average U-value. Table 8 indicates the variation in the area of openings which can be achieved within this constraint.

Summary of provisions in the Elemental method

2.11 Diagram 6 summarises the fabric insulation standards and allowances for windows, doors and rooflights given in the Elemental method. Examples of the procedures used in this method are given in Appendices A, C and E.

Diagram 6 Standard U-values (W/m²K) and areas of openings for buildings other than dwellings

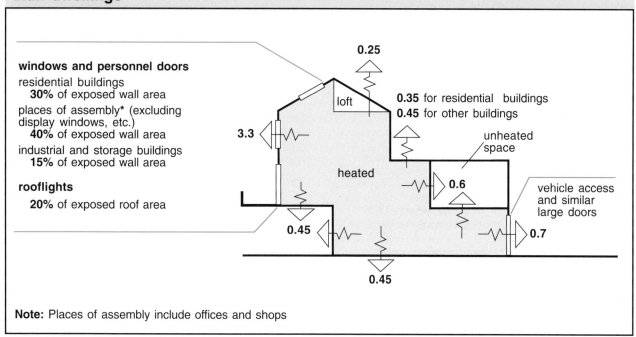

windows and personnel doors
residential buildings
   30% of exposed wall area
places of assembly* (excluding display windows, etc.)
   40% of exposed wall area
industrial and storage buildings
   15% of exposed wall area

rooflights
   20% of exposed roof area

0.25

0.35 for residential buildings
0.45 for other buildings

loft

3.3

unheated space

heated

0.6

vehicle access and similar large doors

0.45

0.45

0.7

Note: Places of assembly include offices and shops

Table 8 Permitted variation in the areas of windows, doors and rooflights for buildings other than dwellings

Average U-value	Residential buildings percentage of wall area	Places of assembly, offices and shops percentage of wall area	Industrial and storage buildings percentage of wall area	Rooflights (all) percentage of roof area
2.0	55	74	28	37
2.1	52	69	26	35
2.2	49	65	24	33
2.3	46	62	23	31
2.4	44	58	22	29
2.5	42	56	21	28
2.6	40	53	20	27
2.7	38	51	19	25
2.8	36	49	18	24
2.9	35	47	17	23
3.0	34	45	17	22
3.1	32	43	16	22
3.2	31	41	16	21
3.3	**30**	**40**	**15**	**20**
3.4	29	39	14	19
3.5	28	37	14	19
3.6	27	36	14	18
3.7	26	35	13	18
3.8	26	34	13	17
3.9	25	33	12	17
4.0	24	32	12	16
4.1	23	31	12	16
4.2	23	30	11	15
4.3	22	30	11	15
4.4	22	29	11	14
4.5	21	28	11	14
4.6	21	27	10	14
4.7	20	27	10	13
4.8	20	26	10	13
4.9	19	26	10	13
5.0	19	25	9	13

Note
The data in this table is derived assuming a constant heat loss through the exposed wall or roof area as appropriate. The constant heat loss amounts to the loss through the wall or roof component plus the loss through the basic area allowance of windows, personnel doors or rooflights respectively as calculated using the U-values in Table 5.

Calculation method

2.12 Within certain limits this method allows greater flexibility than the **Elemental** method in selecting the areas of windows, personnel doors and rooflights and/or the insulation levels of individual elements in the building envelope.

2.13 Calculations should show that the rate of heat loss through the envelope of the proposed building (which could have different U-values or areas of openings from those shown in the **Elemental** method) is not greater than the rate of heat loss from a notional building of the same size and shape designed to comply with the **Elemental** method.

2.14 Larger areas of openings than those shown in Table 6 can be used in the proposed building. If the area of openings in the proposed building is less than that given in Table 6, however, this smaller area should also be assumed in the notional building.

2.15 If the U-value of the floor in the proposed building is better than 0.45 W/m²K with no added insulation the better value should also be assumed in the notional building.

2.16 Examples of the use of the **Calculation** method are included in Appendix H.

Energy Use method

2.17 This is a calculation method allowing completely free design of buildings using any valid energy conservation measure and taking account of useful solar and internal heat gains.

2.18 The requirement will be met if the calculated annual energy use of the proposed building is less than the calculated annual energy use of a similar building designed to comply with the **Elemental** method as set out in paragraphs 2.2 to 2.11 above.

2.19 For buildings which are to be naturally ventilated an acceptable method of demonstrating compliance is given in the CIBSE publication *Building Energy Code* 1981, *Part 2a* (worksheets 1a to 1e).

Using the calculation procedures

Limiting U-values
2.20 When using the **Calculation** method or the **Energy Use** method it may be possible to achieve satisfactory solutions where the U-values of some elements are worse than those set out in Table 5. However, as a general rule the U-values should be limited so that:

a. in residential buildings the roof U-value should not be worse than 0.45 W/m²K and the **exposed** wall and floor U-values should not be worse than 0.7 W/m²K; and

b. in non-residential buildings the U-values of **exposed** walls, roofs and floors should not be worse than 0.7 W/m²K.

Diagram 7 **Reducing thermal bridging around openings**

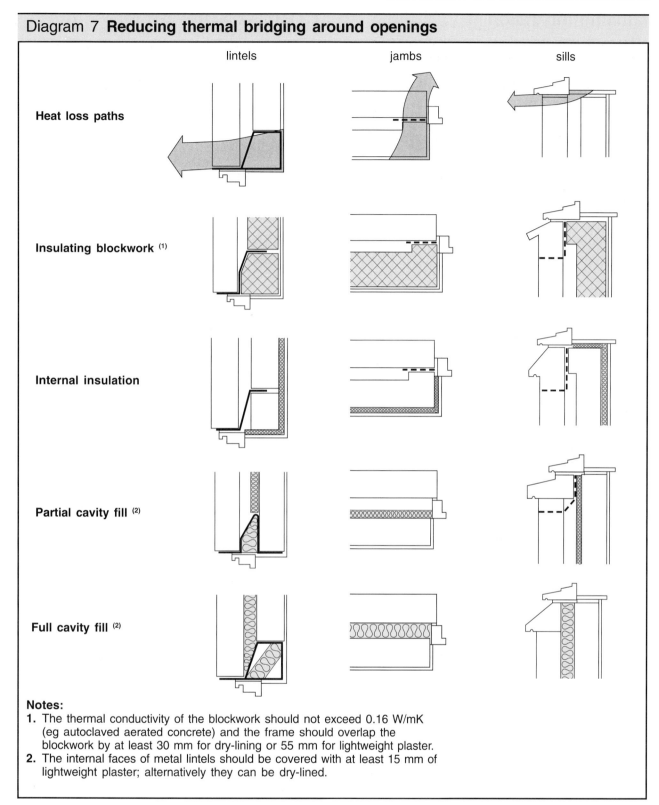

Notes:
1. The thermal conductivity of the blockwork should not exceed 0.16 W/mK (eg autoclaved aerated concrete) and the frame should overlap the blockwork by at least 30 mm for dry-lining or 55 mm for lightweight plaster.
2. The internal faces of metal lintels should be covered with at least 15 mm of lightweight plaster; alternatively they can be dry-lined.

Thermal bridging around openings

2.21   Provision should be made to limit the thermal bridging which occurs around windows, doors and other openings. This is necessary in order to avoid excessive additional heat losses and the possibility of local condensation problems.

2.22   Lintel, jamb and sill designs similar to those shown in Diagram 7 would be satisfactory and the heat losses due to thermal bridging can be ignored if they are adopted.

Alternative method
2.23   An alternative way of demonstrating compliance would be to show by calculation that the edge details around openings will give a satisfactory performance. Appendix D gives a procedure for this.

Limiting infiltration

2.24   Space heating demand is significantly affected by infiltration of cold outside air through leakage paths in the building envelope. It is therefore desirable to limit leakage by reducing unintentional air paths as far as is practicable. One way of satisfying the requirements would be to provide the following measures:

a.   sealing the gaps between dry-lining and masonry walls at the edges of openings such as windows and doors, and at the junctions with walls, floors and ceilings (eg by continuous bands of fixing plaster);

b.   sealing vapour control membranes in timber-frame and other framed-panel constructions;

c.   fitting draught-stripping in the frames of openable elements of windows, doors and rooflights;

d.   sealing around floor and ceiling hatches;

e.   ensuring boxing for concealed services is sealed at floor and ceiling levels and sealing piped services where they penetrate or project into hollow constructions or voids.

2.25   Diagram 8 illustrates sealing measures which could satisfy the requirement. Guidance on methods of reducing infiltration in larger more complex buildings is given in BRE Report BR 265 *Minimising air infiltration in office buildings.*

Diagram 8 **Ways of limiting infiltration in buildings other than dwellings**

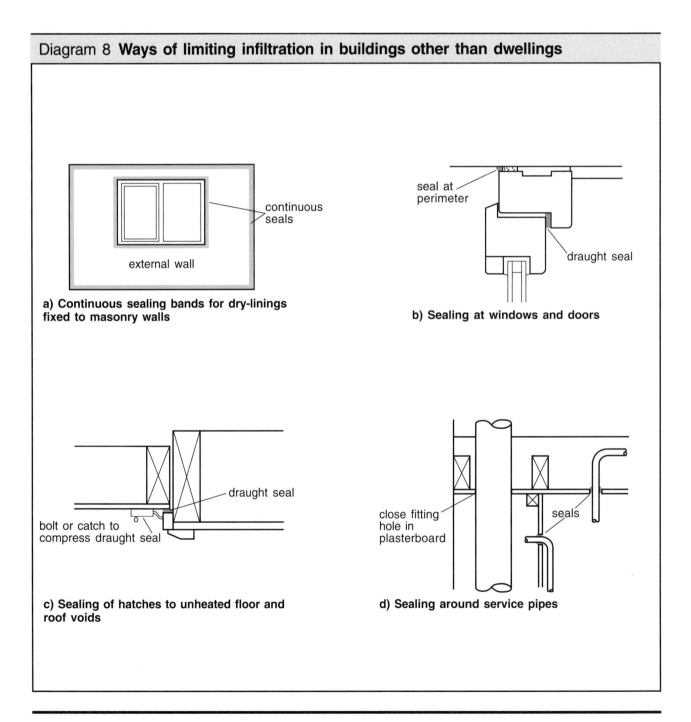

a) Continuous sealing bands for dry-linings fixed to masonry walls

external wall

continuous seals

b) Sealing at windows and doors

seal at perimeter

draught seal

c) Sealing of hatches to unheated floor and roof voids

bolt or catch to compress draught seal

draught seal

d) Sealing around service pipes

close fitting hole in plasterboard

seals

Space heating system controls

2.26 This section is not intended to apply to control systems for commercial or industrial processes.

Temperature controls
2.27 The requirement will be met by the appropriate provision of:

a. thermostats and/or thermostatic radiator valves or any other equivalent form of temperature sensing control for each part of the space heating system designed to be separately controlled; and

b. where the space heating system uses hot water, an external temperature sensor and a weather compensator controller which regulates the temperature of the water flowing in the heating system. See Diagram 9.

Time controls
2.28 The requirement will be met by the provision of heating time controls to maintain the required temperature in each part of the building designed to be separately controlled only when the building is normally occupied. The timing controls could be:

a. for space heating systems with an output of 100 kW or less: clock controls which enable start and stop times to be manually adjusted;

b. for space heating systems with an output in excess of 100 kW: optimising controllers which set the start time for each space heating system depending on the rate at which the building will cool down and then heat up again when the heating is switched off for a period and then re-started.

2.29 Additional controls may be provided to allow sufficient heating for the prevention of condensation or frost damage during periods when the normal heating service would otherwise be switched off.

Boiler sequence controls
2.30 The requirement will be met by the provision of sequence controls for multiple boiler installations where the boilers jointly serve loads in excess of 100 kW. A sequence controller should detect variations in heating demand and start, stop or modulate boilers in combinations which are effective for the purposes of the conservation of fuel and power.

Hot water storage system controls

2.31 For a system other than one heated by a solid fuel fired boiler the requirement will be met if:

a. the **heat exchanger** in the storage vessel has sufficient heating capacity for effective control: a way of satisfying this requirement would be to provide vessels complying with BS 1566 or BS 3198 or equivalent and in particular with the requirements for the surface areas and pipe diameters of heat exchangers given in these standards; and

b. a **thermostat** is provided which shuts off the supply of heat when the storage temperature is reached, and which in the case of hot water central heating systems is interconnected with the room thermostat(s) to switch off the boiler when no heat is required; and

c. a **timer** is provided either as part of the central heating system or as a local device which enables the supply of heat to be shut off for the periods when water heating is not required.

2.32 For systems with solid fuel fired boilers where the cylinder is not providing the slumber load the requirement will be met by the provision of a **thermostatically controlled valve**.

Alternative approaches
2.33 The requirement will be met by adopting the relevant recommendations in the following standards provided they achieve zoning, timing and boiler control performances equivalent to the above:

a. BS 6880: 1988 *Code of practice for low temperature hot water heating systems of output greater than 45 kW;*

b. CIBSE Applications Manual AM1: 1985 *Automatic controls and their implications for systems design.*

Diagram 9 **Outside compensator control for hot water heating systems**

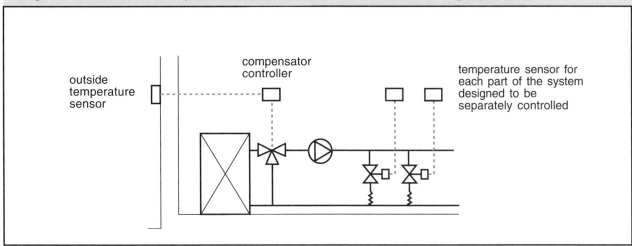

Insulation of vessels, pipes and ducts

Application

2.34   This section is not intended to apply to vessels and piping and ducting systems for commercial and industrial processes.

Insulation of hot water vessels

VESSELS COMPLYING WITH BS 1566 OR BS 3198 OR EQUIVALENT

2.35   For a 120 litre vessel (450 mm dia by 900 mm high) the requirement will be met by providing factory-applied insulation that restricts standing heat losses to 1 W/litre or less in tests using the method in BS 1566 Part 1 Appendix B.4 or equivalent. For other vessel sizes the requirement will be met by providing the same insulation in terms of both material and thickness as required for a 120 litre vessel or equivalent. A way of satisfying the requirement would be to provide vessels with a 35 mm thick, factory-applied coating of PU-foam having zero ozone depletion potential and a minimum density of 30 kg/m .

VESSELS COMPLYING WITH BS 853 OR EQUIVALENT

2.36   The requirement will be met by providing insulation comprising 50 mm of a material having a thermal conductivity of 0.045 W/mK or other material applied in a thickness giving equivalent performance. Note normal practice would include provision of a metal outer casing for physical protection of the insulation.

UNVENTED HOT WATER STORAGE VESSELS

2.37   For unvented hot water systems additional insulation should be provided to control the heat losses through the safety fittings and pipework without impeding safe operation and visibility of warning discharges. See Approved Document G3.

Insulation of pipes and ducts

2.38   Unless the heat loss from a pipe or duct contributes to the useful heat requirement of a room or space the pipe or duct should be insulated. The requirement would be satisfied if:

a.   for pipes the insulation material has a thermal conductivity not exceeding 0.045 W/mK and a thickness equal to the outside diameter of the pipe up to a maximum of 40 mm; OR

b.   for pipes and in the case of warm air ducts insulation is provided in accordance with the recommendations of BS 5422: 1990 *Methods for specifying thermal insulating materials on pipes, ductwork and equipment.*

2.39   The hot pipes connected to hot water storage vessels complying with BS 1566 or BS 3198 or equivalent, including the vent pipe, and the primary flow and return to the heat exchanger where fitted, should be insulated for at least one metre from their points of connection or they should be insulated up to the point where they become concealed. The insulation should comprise 15 mm of a material having a thermal conductivity of 0.045 W/mK or other material applied in a thickness giving equivalent performance.

2.40   The hot pipes connected to hot water storage vessels complying with BS 853 or equivalent including the vent pipe, and the primary and secondary flow and return pipes where fitted, should be insulated in accordance with paragraph 2.38.

2.41   It should be noted that central heating and hot water pipework in unheated areas may need increased insulation thicknesses for the purpose of protection against freezing. Guidance on suitable protection measures is given in BRE Report BR 262 *Thermal insulation: avoiding risks.*

Lighting

General guidance

2.42   For the purposes of the regulations for the conservation of fuel and power:

a.   *emergency escape lighting* means that part of emergency lighting that provides illumination for the safety of people leaving an area or attempting to terminate a dangerous process before leaving an area;

b.   *display lighting* means lighting intended to highlight displays of exhibits or merchandise;

c.   *circuit Watts* means the power consumed by lamps and their associated control gear and power factor correction equipment;

d.   *switch* includes dimmer switches and *switching* includes dimming. As a general rule dimming should be effected by reducing rather than diverting the energy supply.

2.43   The requirements for efficacy and controllability do not apply to display lighting or emergency escape lighting.

Minimum efficacy of lamps

2.44   The requirement will be met if at least 95% of the installed lighting capacity in circuit Watts comprises lighting fittings incorporating lamps of the types listed in Table 9.

Table 9	**Types of high efficacy lamps**
Light source	**Types**
High pressure Sodium	All types and ratings
Metal halide	
Induction lighting	
Tubular fluorescent	All 25mm diameter (T8) lamps provided with low-loss or high frequency control gear
Compact fluorescent	All ratings above 11 W

Provision of lighting controls

2.45 Where it is practical the aim of lighting controls should be to encourage the maximum use of daylight and to avoid unnecessary lighting during the times when spaces are unoccupied. However, the operation of automatically switched lighting systems should not endanger the passage of building occupants.

2.46 Local switches should be provided in easily accessible positions within each working area or at boundaries between working areas and general circulation routes. Local switches could include:

a. switches that are operated by the deliberate action of the occupants either manually or by remote control. Manual switches include rocker switches, press buttons and pull-cords. Remote control switches include infra-red transmitter, sonic, ultra-sonic and telephone handset controllers;

b. automatic switching systems including controls which switch the lighting off when they sense the absence of occupants.

2.47 One way of satisfying the requirement would be the provision of local switching as in Diagram 10 where the distance on plan from any switch to the furthest lighting fitting it controls is generally not more than eight metres or three times the height of the light fitting above floor level if this is greater.

Alternative approaches

Minimum efficacy of lamps

2.48 The requirement would be met if the installed lighting capacity comprises lighting fittings incorporating lamps with an average initial (100 hour) efficacy of not less than 50 lumens per circuit Watt.

Controls in offices and storage buildings

2.49 The requirement would be met by the provision of:

a. local switching for each working area of the building arranged to maximise the beneficial use of daylight. (The local switching provisions would in many cases follow the guidance in paragraphs 2.45 to 2.47); and

b. other controls such as time-switches and photo-electric switches where appropriate.

Controls in other buildings

2.50 The requirement would be met by the provision of one or more of the following types of control system arranged to maximise the beneficial use of daylight as appropriate:

a. local switching as described in paragraphs 2.45 to 2.47 above;

b. time switching: for example, in major operational areas which have clear time-tables of occupation;

c. photo-electric switching.

Using CIBSE guidance

2.51 The requirement would be met by providing lighting controls designed in accordance with the recommendations of the CIBSE publication *Code for interior lighting*. However, in relation to the conservation of fuel and power, the alternative designs should perform no worse than designs which follow the guidance on lighting in this Approved Document.

Diagram 10 **Local lighting controls**

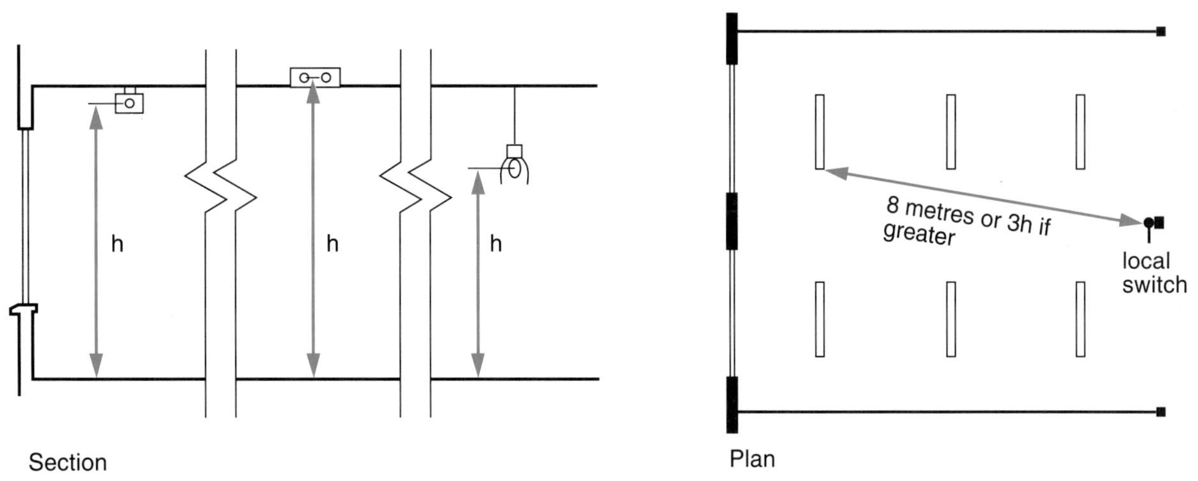

Section

Plan

8 metres or 3h if greater

local switch

Material alterations

2.52 Material alterations are defined in Regulation 3(2) as follows:

An alteration is material for the purposes of these Regulations if the work, or any part of it, would at any stage result –

(a) in a building or controlled service or fitting not complying with a "relevant requirement" where previously it did; or

(b) in a building or controlled service or fitting which before the work commenced did not comply with a "relevant requirement", being more unsatisfactory in relation to such a requirement.

2.53 "Relevant requirement" is defined in Regulation 3(3) as follows:

In paragraph 3(2) "relevant requirement" means any of the following applicable requirements of Schedule 1, namely:

Part A (structure)

paragraph B1 (means of escape)

paragraph B3 (internal fire spread – structure)

paragraph B4 (external fire spread)

paragraph B5 (access and facilities for the fire service)

Part M (access and facilities for disabled people).

2.54 When undertaking a material alteration the requirement may be satisfied in the following ways although the extent of provision will depend upon the circumstances in each case:

a.   roof insulation: when substantially replacing a roof structure – providing insulation to achieve the U-value for new buildings;

b.   floor insulation: where the structure of ground floors is to be substantially replaced – providing insulation in heated rooms to the standard for new buildings;

c.   wall insulation: when substantially replacing complete external walls – providing a reasonable thickness of insulation and incorporating sealing measures as indicated in paragraph 2.24;

d.   when carrying out building work on **space heating and hot water** systems – providing controls and insulation in accordance with paragraphs 2.27 to 2.41 as if they are new installations.

Material changes of use

2.55 Material changes of use are defined in Regulation 5 as follows:

For the purposes of ... these Regulations, there is a material change of use where there is a change in the purposes for which or the circumstances in which a building is used, so that after that change –

(c) the building is used as a hotel or boarding house where previously it was not;

(d) the building is used as an institution where previously it was not;

(e) the building is used as a public building where previously it was not;

(f) the building is not a building described in Classes I to VI in Schedule 2 (ie an exempt building), where previously it was.

2.56 When undertaking a material change of use the requirement may be satisfied in the following ways although the extent of the provision will depend upon the circumstances in each case:

a.   upgrading the insulation in accessible lofts: for example, additional insulation should generally be provided to achieve a U-value not exceeding 0.35 W/m²K where the existing insulation provides a U-value worse than 0.45 W/m²K;

b.   roof insulation: when substantially replacing a roof structure – providing insulation to achieve the U-value for new buildings;

c.   floor insulation: where the structures of ground floors are to be substantially replaced – providing insulation in heated rooms to the standard for new buildings;

d.   wall insulation: when substantially replacing complete external walls – providing a reasonable thickness of insulation and incorporating sealing measures as indicated in paragraph 2.24;

e.   wall insulation: where internal surfaces are to be renovated over a substantial area – upgrading the insulation of exposed and semi-exposed walls. This could be achieved by providing a reasonable thickness of insulating dry-lining sealed in accordance with the guidance in paragraph 2.24;

f.   where **windows** are to be replaced – providing new draught-stripped windows with an average U-value not exceeding 3.3 W/m²K. This could be inappropriate in conservation work and other situations where the existing window design needs to be retained;

g.   when carrying out building work on **space heating and hot water** systems – providing controls and insulation in accordance with paragraphs 2.27 to 2.41 as if they are new installations,

h.   lighting: where lighting systems are to be substantially replaced – providing new lighting systems in accordance with the guidance in paragraphs 2.44 to 2.51.

Appendix A

TABLES FOR THE DETERMINATION OF THE THICKNESSES OF INSULATION REQUIRED TO ACHIEVE GIVEN U-VALUES

Contents of Appendix A

TABLES

Notes

The values in these tables have been derived using the proportional area method, taking into account the effects of thermal bridging where appropriate.

Intermediate values can be obtained from the tables by linear interpolation.

As an alternative to using these tables, the procedures in Appendices B and C can be used to obtain a more accurate calculation of the amount of insulation required.

EXAMPLE CALCULATIONS

Roofs

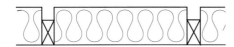

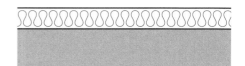

Table A1 **Base thickness of insulation between ceiling joists or rafters**

Thermal conductivity of insulant (W/mK)							
	0.02	0.025	0.03	0.035	0.04	0.045	0.05
Design U-value (W/m²K)	Base thickness of insulating material (mm)						
	A	B	C	D	E	F	G	H
1 0.20	167	209	251	293	335	376	418
2 0.25	114	142	170	199	227	256	284
3 0.30	86	107	129	150	171	193	214
4 0.35	69	86	103	120	137	154	172
5 0.40	57	71	86	100	114	128	143
6 0.45	49	61	73	85	97	110	122

Table A3 **Base thickness for continuous insulation**

Thermal conductivity of insulant (W/mK)							
	0.02	0.025	0.03	0.035	0.04	0.045	0.05
Design U-value (W/m²K)	Base thickness of insulating material (mm)						
	A	B	C	D	E	F	G	H
1 0.20	97	122	146	170	194	219	243
2 0.25	77	97	116	135	154	174	193
3 0.30	64	80	96	112	128	144	160
4 0.35	54	68	82	95	109	122	136
5 0.40	47	59	71	83	94	106	118
6 0.45	42	52	62	73	83	94	104

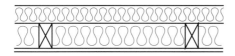

Table A2 **Base thickness of insulation between and over joists or rafters**

Thermal conductivity of insulant (W/mK)							
	0.02	0.025	0.03	0.035	0.04	0.045	0.05
Design U-value (W/m²K)	Base thickness of insulating material (mm)						
	A	B	C	D	E	F	G	H
1 0.20	126	145	166	187	209	232	254
2 0.25	106	120	136	152	169	187	204
3 0.30	86	104	116	129	143	157	171
4 0.35	69	86	102	112	124	135	147
5 0.40	57	71	86	100	109	119	129
6 0.45	49	61	73	85	97	107	115

Table A4 **Allowable reductions in thickness for common roof components**

Thermal conductivity of insulant (W/mK)							
	0.02	0.025	0.03	0.035	0.04	0.045	0.05
Concrete slab density (kg/m³)	Reduction in base thickness of insulating material (mm) for each 100 mm of concrete slab						
	A	B	C	D	E	F	G	H
1 600	11	13	16	18	21	24	26
2 800	9	11	13	15	17	20	22
3 1100	6	7	9	10	12	13	15
4 1300	5	6	7	8	9	10	11
5 1700	3	3	4	5	5	6	7
6 2100	2	2	2	3	3	4	4

Other materials and components	Reduction in base thickness of insulating material (mm)						
7 10 mm plasterboard	1	2	2	2	3	3	3
8 13 mm plasterboard	2	2	2	3	3	4	4
9 13 mm sarking board	2	2	3	3	4	4	5
10 12 mm Calcium Silicate liner board	1	2	2	2	3	3	4
11 Roof space (pitched)	4	5	5	6	7	8	9
12 Roof space (flat)	3	4	5	6	6	7	8
13 19 mm roof tiles	0	1	1	1	1	1	1
14 19 mm asphalt (or 3 layers of felt)	1	1	1	1	2	2	2
15 50 mm screed	2	3	4	4	5	5	6

Note: Tables A1 and A2 are derived for roofs with the proportion of timber at 8%, corresponding to 48 mm wide timbers at 600 mm centres, excluding noggings. For other proportions of timber the U-value can be calculated using the procedure in Appendix B.

Example 1 – Pitched roof with insulation between ceiling joists or between rafters

Determine the thickness of the insulation layer required to achieve a U-value of:
0.25 W/m²K if insulation is between the joists, and **0.35** W/m²K if insulation is between the rafters.

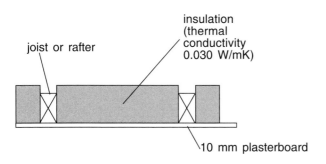

Example 2 – Pitched roof with insulation between and over ceiling joists

Determine the thickness of the insulation layer above the joists required to achieve a U-value of **0.25** W/m²K for the roof construction shown below:

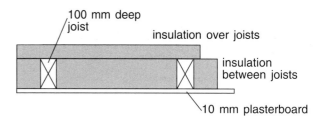

It is proposed to use mineral fibre insulation between and over the joists with a thermal conductivity of 0.035 W/mK.

Using Table A2:

From **column E, row 2** of the table, the base thickness of insulation layer = **152 mm**.

The base thickness may be reduced by taking account of the other materials as follows:

from Table A4:

19 mm roof tiles	column E, row 13	= 1 mm
Roofspace	column E, row 11	= 6 mm
10 mm plasterboard	column E, row 7	= 2 mm
Total reduction		= **9 mm**

The minimum thickness of the insulation layer over the joists required in addition to the 100 mm insulation between the joists to achieve a U-value of 0.25 W/m²K is therefore:

Base thickness less *total reduction*
ie 152 – 100 – 9 = **43 mm**.

For insulation placed between ceiling joists (U-value 0.25 W/m²K)

Using Table A1:

From **column D, row 2** of the table, the base thickness of insulation required is **170 mm**.

The base thickness may be reduced by taking account of the other materials as follows:

from Table A4:

19 mm roof tiles	column D, row 13	= 1 mm
Roofspace	column D, row 11	= 5 mm
10 mm plasterboard	column D, row 7	= 2 mm
Total reduction		= **8 mm**

The minimum thickness of the insulation layer between the ceiling joists required to achieve a U-value of 0.25 W/m²K is therefore:

Base thickness less *total reduction*
ie 170 – 8 = **162 mm**.

For insulation placed between rafters (U-value 0.35 W/m²K)

Using Table A1:

From **column D, row 4** in the table, the base thickness of insulation required is **103 mm**.

The reductions in the base thickness are obtained as follows:

from Table A4:

19 mm roof tiles	column D, row 13	= 1 mm
10 mm plasterboard	column D, row 7	= 2 mm
Total reduction		= **3 mm**

The minimum thickness of the insulation layer between the rafters required to achieve a U-value of 0.35 W/m²K is therefore:

Base thickness less *total reduction*
ie 103 – 3 = **100 mm**.

Approved Document   **Conservation of fuel and power**

27

Example 3 – Industrial roof with outer sheeting

Determine the thickness of the insulating layer required to achieve a U-value of **0.45** W/m²K for the roof construction shown below.

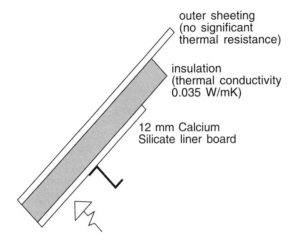

outer sheeting
(no significant
thermal resistance)

insulation
(thermal conductivity
0.035 W/mK)

12 mm Calcium
Silicate liner board

Using Table A3:

From **column E, row 6** of the table, the base thickness of the insulation layer is **73 mm**.

The base thickness may be reduced by taking account of the other materials as follows:

from Table A4:

Outer sheeting (negligible)

12 mm Calcium Silicate **column E, row 10** = 2 mm
liner board

Total reduction from base level thickness   = **2 mm**

The minimum thickness of the insulation layer required to achieve a U-value of 0.45 W/m²K is therefore:

Base thickness less *total reduction*
ie 73 – 2 = **71 mm**.

Example 4 – Concrete deck roof

Determine the thickness of the insulation layer required to achieve a U-value of **0.45** W/m²K for the roof construction shown below.

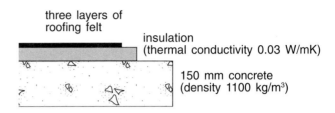

three layers of
roofing felt

insulation
(thermal conductivity 0.03 W/mK)

150 mm concrete
(density 1100 kg/m³)

Using Table A3:

From **column D, row 6** of the table, the base thickness of the insulation layer is **62 mm**.

The base thickness may be reduced by taking account of the other materials as follows:

from Table A4:

3 layers of felt   **column D, row 14** = 1 mm
150 mm concrete deck   **column D, row 3**
adjusted for 150 mm thickness (1.5 x 9)   = 14 mm
Total reduction from base level thickness   = **15 mm**

The minimum thickness of the insulation layer required to achieve a U-value of 0.45 W/m²K is therefore:

Base thickness less *total reduction*
ie 62 – 15 = **47 mm**.

Walls

Walls

Table A5 **Base thickness of insulation layer**

| | | Thermal conductivity of insulant (W/mK) | | | | | | |
| | | 0.02 | 0.025 | 0.03 | 0.035 | 0.04 | 0.045 | 0.05 |
Design U-value (W/m'K)		Base thickness of insulating material (mm)						
	A	B	C	D	E	F	G	H
1	0.30	63	79	95	110	126	142	158
2	0.35	54	67	80	94	107	120	134
3	0.40	46	58	70	81	93	104	116
4	0.45	41	51	61	71	82	92	102
5	0.60	30	37	45	52	59	67	74

Table A6 **Allowable reductions in base thickness for common components**

| | | Thermal conductivity of insulant (W/mK) | | | | | | |
| | | 0.02 | 0.025 | 0.03 | 0.035 | 0.04 | 0.045 | 0.05 |
Component		Reduction in base thickness of insulating material (mm)						
	A	B	C	D	E	F	G	H
1	Cavity (25 mm min.)	4	5	5	6	7	8	9
2	Outer leaf brick	2	3	4	4	5	6	6
3	13 mm plaster	1	1	1	1	1	1	1
4	13 mm lightweight plaster	2	2	2	3	3	4	4
5	10 mm plasterboard	1	2	2	2	3	3	3
6	13 mm plasterboard	2	2	2	3	3	4	4
7	Airspace behind plasterboard dry-lining	2	3	3	4	4	5	6
8	9 mm sheathing ply	1	2	2	2	3	3	3
9	20 mm cement render	1	1	1	1	2	2	2
10	13 mm tile hanging	0	0	0	1	1	1	1

Table A7 **Allowable reduction in base thickness for concrete components**

| | | Thermal conductivity of insulant (W/mK) | | | | | | |
| | | 0.02 | 0.025 | 0.03 | 0.035 | 0.04 | 0.045 | 0.05 |
Density (kg/m)		Reduction in base thickness of insulation (mm) for each 100 mm of concrete						
	A	B	C	D	E	F	G	H
Concrete inner leaf								
1	600	9	11	13	15	17	20	22
2	800	7	9	11	13	15	17	19
3	1000	6	8	9	11	12	14	15
4	1200	5	6	7	9	10	11	12
5	1400	4	5	6	7	8	9	9
6	1600	3	4	4	5	6	7	7
Concrete outer leaf or single leaf wall								
7	600	8	10	13	15	17	19	21
8	800	7	8	10	12	14	15	17
9	1000	6	7	8	10	11	12	14
10	1200	4	6	7	8	9	10	11
11	1400	3	4	5	6	7	8	9
12	1600	3	3	4	5	5	6	7
13	1800	2	3	3	4	4	5	5
14	2000	2	2	2	3	3	4	4
15	2400	1	1	2	2	2	2	3

Table A8 **Allowable reduction in base thickness for insulated timber frame walls**

| | | Thermal conductivity of insulant (W/mK) | | | | | | |
| | | 0.02 | 0.025 | 0.03 | 0.035 | 0.04 | 0.045 | 0.05 |
Thermal conductivity of insulation within frame (W/mK)		Reduction in base thickness of insulation for each 100 mm of frame (mm)						
	A	B	C	D	E	F	G	H
1	0.035		42	53	63	74	84	95	105
2	0.040		38	48	58	67	77	87	96

Note: The table is derived for walls for which the proportion of timber is 12%, which corresponds to 48 mm wide studs at 400 mm centres. For other proportions of timber the U-value can be calculated using the procedure in Appendix B.

Example 5 – Masonry cavity wall with internal insulation

Determine the thickness of the insulation layer required to achieve a U-value of **0.45** W/m²K for the wall construction shown below.

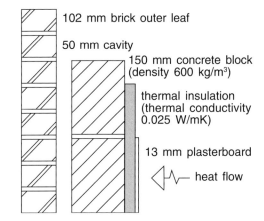

- 102 mm brick outer leaf
- 50 mm cavity
- 150 mm concrete block (density 600 kg/m³)
- thermal insulation (thermal conductivity 0.025 W/mK)
- 13 mm plasterboard
- heat flow

Using Table A5:

From column C, row 4 of the table, the base thickness of the insulation layer is **51 mm**.

The base thickness may be reduced by taking account of the other materials as follows:

from Table A6:

Brick outer leaf	column C, row 2	= 3 mm
Cavity	column C, row 1	= 5 mm
Plasterboard	column C, row 6	= 2 mm

and from Table A7:

Concrete block	column C, row 1	
adjusted for 150 mm block thickness (1.5 x 11)		=17 mm
Total reduction		**=27 mm**

The minimum thickness of the insulation layer required to achieve a U-value of 0.45 W/m²K is therefore:

Base thickness less *total reduction*
ie 51 – 27 = **24 mm.**

Example 6 – Masonry cavity wall filled with insulation with plasterboard on dabs

Determine the thickness of the insulation layer required to achieve a U-value of **0.45** W/m²K for the wall construction shown below.

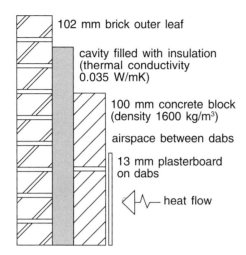

102 mm brick outer leaf

cavity filled with insulation (thermal conductivity 0.035 W/mK)

100 mm concrete block (density 1600 kg/m³)

airspace between dabs

13 mm plasterboard on dabs

heat flow

Using Table A5:

From column E, row 4 of the table, the base thickness of the insulation layer is **71 mm**.

The base thickness may be reduced by taking account of the other materials as follows:

from Table A6:

Brick outer leaf	column E, row 2	=	4 mm
13 mm plasterboard	column E, row 6	=	3 mm
Airspace behind plasterboard			
	column E, row 7	=	4 mm

and from Table A7:

Concrete block	column E, row 6	=	5 mm
Total reduction		=	**16 mm**

The minimum thickness of the insulation layer required to achieve a U-value of 0.45 W/m²K is therefore:

Base thickness less *total reduction*
ie 71 – 16 = **55 mm**.

Example 7 – Masonry wall with partial cavity-fill

Determine the thickness of the insulation layer required to achieve a U-value of **0.45** W/m²K for the wall construction shown below.

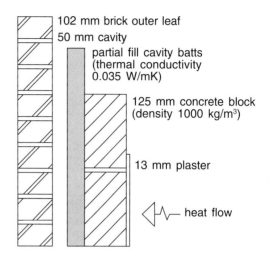

102 mm brick outer leaf

50 mm cavity

partial fill cavity batts (thermal conductivity 0.035 W/mK)

125 mm concrete block (density 1000 kg/m³)

13 mm plaster

heat flow

Using Table A5:

From column E, row 4 of the table, the base thickness of the insulation layer is **71 mm**.

The base thickness may be reduced by taking account of the other materials as follows:

from Table A6:

Brick outer leaf	column E, row 2	=	4 mm
Cavity	column E, row 1	=	6 mm
Plaster	column E, row 3	=	1 mm

and from Table A7:

Concrete block	column F, row 3		
adjusted for 125 mm thickness (1.25 x 11)		=	14 mm
Total reduction		=	**25 mm**

The minimum thickness of the insulation layer required to achieve a U-value of 0.45 Wm²K is therefore:

Base thickness less *total reduction*
ie 71 – 25 = **46 mm.**

Example 8 – Semi-exposed solid wall

Determine the thickness of the insulation layer required to achieve a U-value of **0.6** W/m²K for the wall construction shown below.

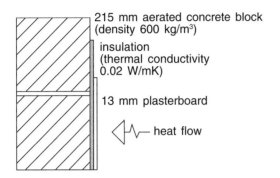

215 mm aerated concrete block (density 600 kg/m³)

insulation (thermal conductivity 0.02 W/mK)

13 mm plasterboard

heat flow

Using Table A5:

From column B, row 5 of the table, the base thickness of the insulation layer is **30 mm.**

The base thickness may be reduced by taking account of the other materials as follows:

from Table A6:

Plasterboard	column B, row 6	= 2 mm

and from Table A7:

Concrete block adjusted for 215 mm thickness (2.15 x 8)	column B, row 7	= 17 mm
Total reduction		= **19 mm**

The minimum thickness of the insulation layer required to achieve a U-value of 0.6 W/m²K is therefore:

Base thickness less *total reduction*
ie 30 – 19 = **11 mm**.

Example 9 – Timber-frame wall

Determine the thickness of the insulation layer required to achieve a U-value of **0.45** W/m²K for the wall construction shown below.

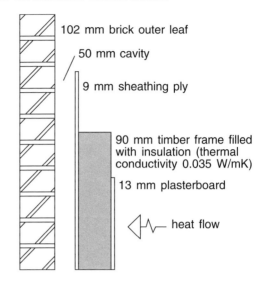

102 mm brick outer leaf

50 mm cavity

9 mm sheathing ply

90 mm timber frame filled with insulation (thermal conductivity 0.035 W/mK)

13 mm plasterboard

heat flow

Using Table A5:

From column E, row 4 of the table, the base thickness of the insulation layer is **71 mm**.

The base thickness may be reduced by taking account of the other materials as follows:

from Table A6:

Brick outer leaf	column E, row 2	= 4 mm
Cavity	column E, row 1	= 6 mm
Sheathing ply	column E, row 8	= 2 mm
Plasterboard	column E, row 6	= 3 mm

and from Table A8:

Timber frame adjusted for shallower member (0.9 x 74mm)	column E, row 1	= 67 mm
Total reduction		= **82 mm**

The reduction in base thickness is greater than the required thickness therefore no additional insulation is required.

The builder may wish to substitute a cheaper insulant with conductivity of 0.04 W/mK. In this case, the base thickness of insulation from Table A5 column F, row 4 would be **82 mm** but the reductions due to the effects of the other materials would equal this and no additional insulation need be provided.

However, the builder may wish to increase the standard of insulation **by reducing the U-value to 0.35 W/m²K**. The base thickness of insulation from Table A5 column E, row 2 would be **94 mm**. The total reduction in base thickness is as before, ie 82 mm and so the minimum thickness of insulation required in addition to the insulation between frame members to achieve a U-value of 0.35 W/m²K is therefore:

Base thickness less *total reduction*
ie 94 – 82 = **12 mm**.

Floors

Ground floors

Note: in using the tables for floors it is first necessary to calculate the ratio P/A, where:

P is the floor perimeter length in metres and

A is the floor area in square metres.

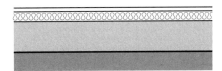

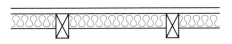

Table A9 **Insulation thickness for solid floors in contact with the ground**

	P/A *	0.02	0.025	0.03	0.035	0.04	0.045	0.05
	A	B	C	D	E	F	G	H

Insulation thickness (mm) for: U-value of 0.25 W/m²K — Thermal conductivity of insulant (W/mK)

	P/A *	0.02	0.025	0.03	0.035	0.04	0.045	0.05
	A	B	C	D	E	F	G	H
1	1.00	62	77	93	108	124	139	155
2	0.90	61	76	91	107	122	137	152
3	0.80	60	75	90	105	119	134	149
4	0.70	58	73	87	102	116	131	145
5	0.60	56	70	84	98	111	125	139
6	0.50	52	66	79	92	105	118	131
7	0.40	47	59	71	83	95	107	119
8	0.30	39	49	59	69	79	88	98
9	0.20	24	30	36	42	48	54	60

U-value of 0.35 W/m²K

10	1.00	39	49	58	68	78	88	97
11	0.90	38	48	57	67	76	86	95
12	0.80	37	46	55	65	74	83	92
13	0.70	35	44	53	62	70	79	88
14	0.60	33	41	49	58	66	74	82
15	0.50	30	37	44	52	59	67	74
16	0.40	25	31	37	43	49	55	61
17	0.30	16	21	25	29	33	37	41
18	0.20	1	1	1	2	2	2	2

U-value of 0.45 W/m²K

19	1.00	26	33	39	46	53	59	66
20	0.90	25	32	38	44	51	57	63
21	0.80	24	30	36	42	48	54	60
22	0.70	22	28	34	39	45	51	56
23	0.60	20	25	30	35	40	45	50
24	0.50	17	21	25	30	34	38	42
25	0.40	12	15	18	21	24	27	30
26	0.30	4	5	6	6	7	8	9
27	<0.27	0	0	0	0	0	0	0

Note: * P/A is the ratio of floor perimeter (m) to floor area (m²).

Table A10 **Insulation thickness for suspended timber ground floors**

Insulation thickness (mm) for: U-value of 0.25 W/m²K — Thermal conductivity of insulant (W/mK)

	P/A *	0.02	0.025	0.03	0.035	0.04	0.045	0.05
	A	B	C	D	E	F	G	H
1	1.00	95	110	126	140	155	170	184
2	0.90	93	109	124	138	153	167	181
3	0.80	91	106	121	136	150	164	178
4	0.70	88	103	118	132	145	159	173
5	0.60	85	99	113	126	139	153	166
6	0.50	79	92	106	118	131	143	156
7	0.40	71	83	95	107	118	129	140
8	0.30	57	68	78	88	97	106	116
9	0.20	33	39	46	52	58	64	69

U-value of 0.35 W/m²K

10	1.00	57	67	77	87	96	106	115
11	0.90	55	66	75	85	94	103	112
12	0.80	53	63	73	82	91	100	109
13	0.70	51	60	69	78	87	95	104
14	0.60	47	56	64	73	81	89	97
15	0.50	42	50	57	65	72	79	87
16	0.40	34	41	47	53	60	66	72
17	0.30	22	26	31	35	39	43	47
18	0.20	1	1	2	2	2	2	3

U-value of 0.45 W/m²K

19	1.00	37	44	51	57	64	70	77
20	0.90	35	42	49	55	62	68	74
21	0.80	33	40	46	53	59	65	70
22	0.70	31	37	43	49	54	60	65
23	0.60	27	33	38	43	49	54	58
24	0.50	22	27	32	36	40	44	49
25	0.40	15	18	22	25	28	31	34
26	0.30	4	5	6	7	8	9	10
27	<0.27	0	0	0	0	0	0	0

Notes: * P/A is the ratio of floor perimeter (m) to floor area (m²).

The table is derived for suspended timber floors for which the proportion of timber is 12%, which corresponds to 48 mm wide timbers at 400 mm centres.

For other proportions of timber the U-value can be calculated using the procedure in Appendix B.

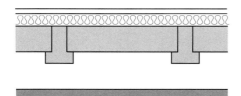

Table A11 Insulation thickness for suspended concrete beam and block ground floors

		Insulation thickness (mm) for: U-value of 0.25 W/m²K						
		Thermal conductivity of insulant (W/mK)						
	P/A *	0.02	0.025	0.03	0.035	0.04	0.045	0.05
	A	B	C	D	E	F	G	H
1	1.00	60	75	90	104	119	134	149
2	0.90	59	73	88	103	118	132	147
3	0.80	58	72	86	101	115	130	144
4	0.70	56	70	84	98	112	126	140
5	0.60	54	67	80	94	107	121	134
6	0.50	50	63	75	88	101	113	126
7	0.40	45	57	68	79	91	102	113
8	0.30	37	46	56	65	74	84	93
9	0.20	22	27	33	38	43	49	54
		U-value of 0.35 W/m²K						
10	1.00	37	46	55	64	74	83	92
11	0.90	36	45	54	63	72	81	90
12	0.80	35	43	52	61	69	78	87
13	0.70	33	41	50	58	66	74	83
14	0.60	31	38	46	54	61	69	77
15	0.50	27	34	41	48	55	62	69
16	0.40	22	28	34	39	45	50	56
17	0.30	14	18	21	25	29	32	36
18	0.20	0	0	0	0	0	0	0
		U-value of 0.45 W/m²K						
19	1.00	24	30	36	42	48	54	60
20	0.90	23	29	35	41	46	52	58
21	0.80	22	28	33	39	44	50	55
22	0.70	20	25	31	36	41	46	51
23	0.60	18	23	27	32	36	41	45
24	0.50	15	18	22	26	29	33	37
25	0.40	10	12	15	17	19	22	24
26	0.30	2	2	2	3	3	4	4
27	< 0.28	0	0	0	0	0	0	0

Note: * P/A is the ratio of floor perimeter (m) to floor area (m²).

Example 10 – Solid floor in contact with the ground

Determine the thickness of the insulation layer required to achieve a U-value of **0.45** W/m²K for the ground floor slab shown below.

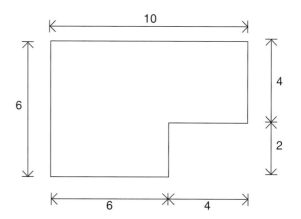

It is proposed to use insulation with a thermal conductivity of 0.02 W/mK.

The overall perimeter length of the slab is:
(10 + 4 + 4 + 2 + 6 + 6) = 32 m

The *floor area* of the slab is:
(6 x 6) + (4 x 4)   = 52 m²

The ratio:
$$\frac{\text{perimeter length}}{\text{floor area}} = \frac{32}{52} = 0.6$$

Using Table A9, column B, row 23 indicates **20 mm** of insulation is required.

The insulation may be located above or below the concrete slab.

Example 11 – Suspended timber floor

If the floor shown above was of suspended timber construction, the perimeter length and floor area would be the same, yielding the same ratio of:

$$\frac{\text{perimeter length}}{\text{floor area}} = \frac{32}{52} = 0.6$$

To achieve a U-value of **0.45** W/m²K, using insulation with a thermal conductivity of 0.04 W/mK, Table A10 column F, row 23 indicates that the insulation thickness between the joists should be not less than **49 mm**.

Exposed and semi-exposed upper floors

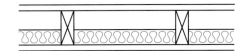

Table A12 Exposed and semi-exposed upper floors of timber construction

Thermal conductivity of insulant (W/mK)							
	0.02	0.025	0.03	0.035	0.04	0.045	0.05
Design U-value (W/m²K)	Base thickness of insulation between joists to achieve design U-values						
A	B	C	D	E	F	G	H
Exposed floor							
1 0.35	61	76	92	107	122	146	162
2 0.45	42	53	63	74	84	95	106
Semi-exposed floor							
3 0.60	25	32	38	44	50	57	63

Note: Table A12 is derived for floors with the proportion of timber at 12% which corresponds to 48 mm wide timbers at 400 mm centres. For other proportions of timber the U-value can be calculated using the procedure in Appendix E.

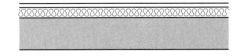

Table A13 Exposed and semi-exposed upper floors of concrete construction

Thermal conductivity of insulant (W/mK)							
	0.02	0.025	0.03	0.035	0.04	0.045	0.05
Design U-value (W/m²K)	Base thickness of insulation to achieve design U-values						
A	B	C	D	E	F	G	H
Exposed floor							
1 0.35	52	65	78	91	104	117	130
2 0.45	39	49	59	69	79	89	98
Semi-exposed floor							
3 0.60	26	33	39	46	52	59	65

Table A14 Exposed and semi-exposed upper floors: allowable reductions in base thickness for common components

Thermal conductivity of insulant (W/mK)							
	0.02	0.025	0.03	0.035	0.04	0.045	0.05
Component	Reduction in base thickness of insulating material (mm)						
A	B	C	D	E	F	G	H
1 10 mm plasterboard	1	2	2	2	3	3	3
2 19 mm timber flooring	3	3	4	5	5	6	7
3 50 mm screed	2	3	4	4	5	5	6

Table A15 Thermal conductivity of some common building materials

Material	Density (kg/m³)	Thermal conductivity (W/mK)
Walls (external and internal)		
Brickwork (outer leaf)	1700	0.84
Brickwork (inner leaf)	1700	0.62
Cast concrete (dense)	2100	1.40
Cast concrete (lightweight)	1200	0.38
Concrete block (heavyweight)	2300	1.63
Concrete block (medium weight)	1400	0.51
Concrete block (lightweight)	600	0.19
Normal mortar	1750	0.8
Fibreboard	300	0.06
Plasterboard	950	0.16
Tile hanging	1900	0.84
Timber	650	0.14
Surface finishes		
External rendering	1300	0.50
Plaster (dense)	1300	0.50
Plaster (lightweight)	600	0.16
Calcium Silicate board	875	0.17
Roofs		
Aerated concrete slab	500	0.16
Asphalt	1700	0.50
Felt/bitumen layers	1700	0.50
Screed	1200	0.41
Stone chippings	1800	0.96
Tile	1900	0.84
Wood wool slab	500	0.10
Floors		
Cast concrete	2000	1.13
Metal tray	7800	50.00
Screed	1200	0.41
Timber flooring	650	0.14
Wood blocks	650	0.14
Insulation		
Expanded polystyrene (EPS) slab	25	0.035
Mineral wool quilt	12	0.040
Mineral wool slab	25	0.035
Phenolic foam board	30	0.020
Polyurethane board	30	0.025

Note: If available, certified test values should be used in preference to those in the table.

Appendix B

THE PROPORTIONAL AREA CALCULATION METHOD FOR DETERMINING U-VALUES OF STRUCTURES CONTAINING REPEATING THERMAL BRIDGES

B1 Full details of the calculation method are given in the CIBSE Design Guide Section A3. However, two examples are given in this Appendix which illustrate the method as applied to frequently encountered designs.

B2 If the element design does not include a continuous cavity all the element layers have to be analysed together. Where the design does incorporate a continuous cavity, however, the section should be divided into two parts along the centre of the cavity and the parts analysed separately. The resistances of the two parts should be determined with half of the cavity resistance assigned to each. The thermal resistances can then be added together to obtain the total resistance of the element.

B3 In some cases, the joists in timber roof and floor constructions will project beyond the surface of the insulation. For the purposes of demonstrating compliance with the requirements for the conservation of fuel and power, however, the calculations should take the depths of the joists to be the same as the depth of insulation, hence ignoring the effect of the projections. Joists which are wholly beneath insulation can also be ignored.

Example 1

What is the U-value of the proposed wall construction shown below?

102 mm brick, thermal conductivity 0.84 W/mK

50 mm cavity, thermal resistance 0.18m²K/W

9 mm sheathing ply, thermal conductivity 0.14 W/mK

90 x 38 mm timber studs at 600 mm centres, thermal conductivity 0.14 W/mK, with insulation, thermal conductivity 0.04 W/mK, between the studs

13 mm plasterboard, thermal conductivity 0.16 W/mK

The resistance of the outside surface of the wall is 0.06 m²K/W and the inside surface resistance is 0.12 m²K/W.

With this construction the thermal bridging of the insulation by the timber studs must be taken into account as follows:

Consider the wall as inner and outer leaves with the boundary between leaves at the centre of the cavity.

Resistance of inner leaf
Resistance through section containing timber stud:

Inside surface resistance	= 0.12 m²K/W
Resistance of plasterboard = 0.013/0.16	= 0.08 m²K/W
Resistance of timber stud = 0.09/0.14	= 0.64 m²K/W
Resistance of sheathing ply = 0.009/0.14	= 0.06 m²K/W
Half cavity resistance	= 0.09 m²K/W
Resistance of section through timber stud:	R_t = **0.99 m²K/W**

Resistance through section containing insulation:

Inside surface resistance	= 0.12 m²K/W
Resistance of plasterboard = 0.013/0.16	= 0.08 m²K/W
Resistance of insulation = 0.09/0.04	= 2.25 m²K/W
Resistance of sheathing ply = 0.009/0.14	= 0.06 m²K/W
Half cavity resistance	= 0.09 m²K/W
Resistance of section through insulation:	R_{ins} = **2.60 m²K/W**

Combination of these resistances:

Fractional area of timber stud:

$$F_t = \frac{\text{thickness of studs}}{\text{stud centres}} = \frac{38}{600} = 0.063$$

Fractional area of insulation: $F_{ins} = (1 - F_t)$ = 0.937

The resistance of the inner leaf is then obtained from:

$$R_{inner} = \frac{1}{\dfrac{F_t}{R_t} + \dfrac{F_{ins}}{R_{ins}}} = \frac{1}{\dfrac{0.063}{0.99} + \dfrac{0.937}{2.60}} = \textbf{2.36 m}^2\textbf{K/W}$$

Resistance of outer leaf
This is treated as an unbridged structure (the difference in resistance between brick and mortar is less than 0.1 m″K/W).

Half cavity resistance	= 0.09 m²K/W
Resistance of brick = 0.102/0.84	= 0.12 m²K/W
Outside surface resistance	= 0.06 m²K/W
Resistance of outer leaf: R_{outer}	= **0.27 m²K/W**

Total resistance of wall
The total resistance of the wall is the sum of the resistances of the inner and outer leaves:

$R_{inner} + R_{outer}$ = 2.36 + 0.27

Hence total resistance = **2.63 m²K/W**

U-value of wall
The U-value is given by: $U = \dfrac{1}{\text{total resistance}}$

So: $U = \dfrac{1}{2.63}$ = **0.38 W/m²K**

Example 2

If the proposed wall construction is now that shown below, there are two thermally bridged layers:

a) that of the blockwork by the normal mortar joints, and

b) that of the insulation by the timber studs.

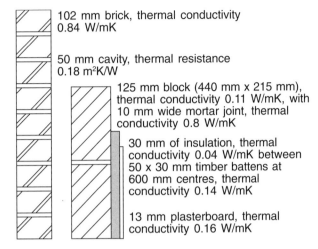

102 mm brick, thermal conductivity 0.84 W/mK

50 mm cavity, thermal resistance 0.18 m²K/W

125 mm block (440 mm x 215 mm), thermal conductivity 0.11 W/mK, with 10 mm wide mortar joint, thermal conductivity 0.8 W/mK

30 mm of insulation, thermal conductivity 0.04 W/mK between 50 x 30 mm timber battens at 600 mm centres, thermal conductivity 0.14 W/mK

13 mm plasterboard, thermal conductivity 0.16 W/mK

Consider the wall as inner and outer leaves with the boundary between leaves as half way through the cavity.

Resistance of inner leaf

In this case there are two bridged layers and the precise location of the bridge in one layer with respect to the bridge in another layer is generally unknown (or is not readily determined). There are four different combinations of paths through the blockwork and insulation and it is therefore assumed, because pitch centres do not coincide, that heat flows through them in proportion to their relative areas. The average resistance of the leaf is determined as follows:

RESISTANCE OF NON-BRIDGED LAYERS

Resistance of half the cavity	= 0.09 m²K/W
Resistance of plasterboard = 0.013/0.16	= 0.08 m²K/W
Resistance of inside surface	= 0.12 m²K/W
Resistance of non-bridged layers: R_{nb}	**= 0.29 m²K/W**

RESISTANCE OF BRIDGED LAYERS
Heat flow paths

The two bridged layers create four paths:

block/insulation

block/timber

mortar/insulation

mortar/timber.

Material resistances

Resistance of block:	$R_b = 0.125/0.11$	$= 1.14$ m²K/W
Resistance of mortar:	$R_m = 0.125/0.8$	$= 0.16$ m²K/W
Resistance of insulation:	$R_{ins} = 0.03/0.04$	$= 0.75$ m²K/W
Resistance of timber:	$R_t = 0.03/0.14$	$= 0.21$ m²K/W

Resistance of heat flow paths

Resistance of block/insulation:
$$R_{b,ins} = R_b + R_{ins} + R_{nb} \qquad = 2.18 \text{ m}^2\text{K/W}$$

block/timber:
$$R_{b,t} = R_b + R_t + R_{nb} \qquad = 1.64 \text{ m}^2\text{K/W}$$

mortar/insulation:
$$R_{m,ins} = R_m + R_{ins} + R_{nb} \qquad = 1.20 \text{ m}^2\text{K/W}$$

mortar/timber:
$$R_{m,t} = R_m + R_t + R_{nb} \qquad = 0.66 \text{ m}^2\text{K/W}$$

Fraction of face area of materials

block:	$F_b = \dfrac{440 \times 215}{450 \times 225}$	$= 0.934$
mortar:	$F_m = 1 - F_b$	$= 0.066$
insulation:	$F_{ins} = \dfrac{550}{600}$	$= 0.917$
timber:	$F_t = 1 - F_{ins}$	$= 0.083$

Fraction of face area of heat flow paths

block/insulation:	$F_{b,ins} = F_b \times F_{ins}$	$= 0.856$
block/timber:	$F_{b,t} = F_b \times F_t$	$= 0.078$
mortar/insulation:	$F_{m,ins} = F_m \times F_{ins}$	$= 0.061$
mortar/timber:	$F_{m,t} = F_m \times F_t$	$= 0.005$

Sum of parallel resistances

The sum of resistances in parallel is given by the formula:

$$\frac{1}{R_{inner\ leaf}} = \frac{F_{b,ins}}{R_{b,ins}} + \frac{F_{b,t}}{R_{b,t}} + \frac{F_{m,ins}}{R_{m,ins}} + \frac{F_{m,t}}{R_{m,t}}$$

The figures in this case are:

$$\frac{1}{R_{inner\ leaf}} = \frac{0.856}{2.18} + \frac{0.078}{1.64} + \frac{0.061}{1.20} + \frac{0.005}{0.66} = 0.499$$

The resistance of the inner leaf is therefore:

$$R_{inner\ leaf} = \frac{1}{0.499} = \textbf{2.0 m}^2\textbf{K/W}$$

Resistance of outer leaf

Resistance of outside surface	= 0.06 m²K/W
Resistance of brick outer leaf = 0.102/0.84	= 0.12 m²K/W
Resistance of half the cavity	= 0.09 m²K/W
Resistance of outer leaf: $R_{outer\ leaf}$	**= 0.27 m²K/W**

Total resistance of the wall

The total resistance of the wall is the sum of the inner and outer leaf resistances:

$$R_{total} = R_{inner\ leaf} + R_{outer\ leaf} = 2.0 + 0.27 \qquad \textbf{= 2.27 m}^2\textbf{K/W}$$

U-value of the wall

The wall U-value is given by:

$$U = \frac{1}{R_{total}} = \frac{1}{2.27} = \textbf{0.44 W/m}^2\textbf{K}$$

Appendix C

CALCULATION OF U-VALUES OF GROUND FLOORS

General

C1 A ground floor may have a U-value of 0.45 W/m²K or 0.35 W/m²K without additional insulation if the floor is sufficiently large. Diagram C1 shows the range of floor dimensions for which insulation is required.

Diagram C1 **Floor dimensions for which insulation is required**

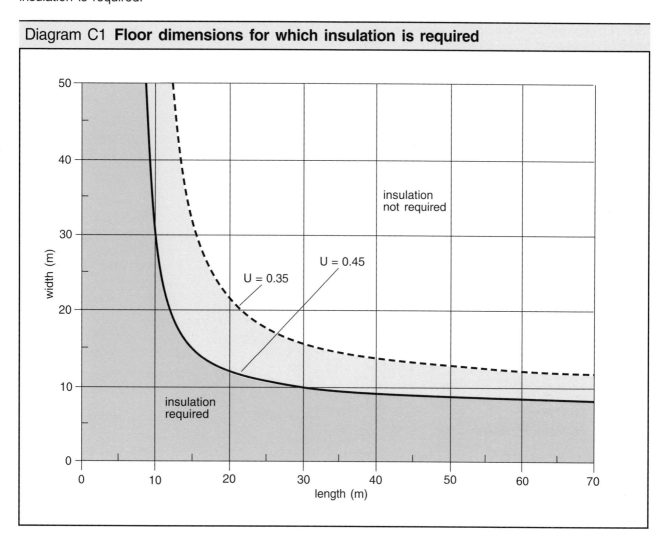

C2 Floor dimensions should be measured in accordance with the convention described in paragraph 0.12. In the case of semi-detached or terraced premises, blocks of flats and similar, the floor dimensions can either be taken as those of the premises themselves, or of the whole building. When considering extensions to existing buildings the floor dimensions can be taken as those of the complete building including the extension.

C3 Care should be taken to control the risk of condensation caused by thermal bridging at the floor edge. See BRE Report BR 262 *Thermal insulation: avoiding risks.*

Determining the U-value of floors with no insulation

C4 The U-value of an uninsulated floor may be determined from the ratio of its exposed perimeter to its area, using the equation:

$$U_o = 0.05 + 1.65 \left(\frac{P}{A}\right) - 0.6 \left(\frac{P}{A}\right)^2$$

where:

U_o = U-value of uninsulated floor (W/m²K)

P = exposed perimeter of floor (m)

A = area of floor (m²)

C5 The equation in paragraph C4 applies to all types of uninsulated floors constructed next to the ground including slab-on-ground, concrete raft, suspended timber and beam-and-block.

C6 Unheated spaces outside the insulated fabric, such as attached garages or porches, should be excluded when determining **P** and **A** but the length of the wall between the heated building and the unheated space should be included when determining the perimeter.

C7 The data in Table C1 has been derived from the equation in paragraph C4. For the purposes of regulations for the conservation of fuel and power it will be sufficient to use the table using linear interpolation where necessary.

Table C1 U-values of uninsulated floors

Ratio P/A	U_o
0.1	0.21
0.2	0.36
0.3	0.49
0.4	0.61
0.5	0.73
0.6	0.82
0.7	0.91
0.8	0.99
0.9	1.05
1.0	1.10

U-value of insulated floor

C8 The U-value of an insulated floor is obtained from:

$$U_{ins} = \frac{1}{\dfrac{1}{U_o} + R_{ins}}$$

where:

R_{ins} is the thermal resistance of the insulation, and

U_o is obtained from Table C1 or the equation in paragraph C4.

C9 In the case of suspended floors U_{ins} includes the thermal resistance of the structural deck and R_{ins} should only include:

a) the resistance of added insulation layers; and/or

b) any extra resistance of the structural deck over and above 0.2 m²K/W.

C10 For further information on floor U-values see BRE IP 3/90. BRE IP 7/93 shows how the U-value of a floor is modified by edge insulation (including low-density foundations), and BRE IP 14/94 gives procedures for basements.

Appendix D

THERMAL BRIDGING AT THE EDGES OF OPENINGS

Summary
D1 As an alternative to the examples given in Diagrams 4 and 9, this Appendix gives a procedure for establishing whether:

a. there is an unacceptable risk of condensation at the edges of openings; and/or

b. the heat losses at the edges of openings are significant.

D2 The procedure involves the assessment of the minimum thermal resistance between inside and outside surfaces at the edges of openings. This requires identification of minimum thermal resistance paths, and calculation of their thermal resistance, taking into account the effect of thin layers such as metal lintels.

D3 These minimum thermal resistances are then compared with satisfactory performance criteria to see whether corrective action is indicated.

Minimum thermal resistance path
D4 The minimum thermal resistance path through a thermal bridge is that path from internal surface to external surface which has the smallest thermal resistance, R_{min}. Diagram D1 illustrates this for a section through a window jamb.

Diagram D1 **Minimum thermal resistance path**

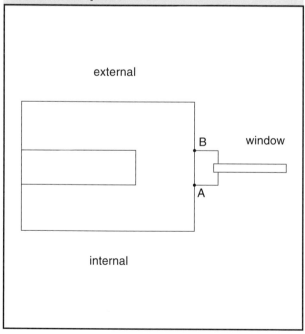

The minimum resistance path in this case is from the internal surface at A to the external surface at B.

R_{min} is equal to the total length from inside to outside (AB) divided by the thermal conductivity of the material of the jamb. An example calculation is given on the following page.

Additional calculation for thin layers such as metal lintels
D5 For details containing thin layers of thickness not exceeding 4 mm (such as metal lintels), a second modified calculation of minimum thermal resistance (R_{mod}) is made wherein the effective thermal conductivity of the thin layer is taken as the largest of 0.1 W/mK or the thermal conductivities of the materials immediately on either side of it. An example of this more complex calculation is given in BRE IP 12/94 *Assessing condensation risk and heat loss at thermal bridges around openings.*

Risk of surface condensation
D6 The risk of surface condensation and mould growth at the edges of openings can be assumed to be negligible if:

a. for edges containing thin layers of thickness not exceeding 4 mm:

R_{min} (rounded to two decimal places) is at least 0.10 m²K/W; and

R_{mod} (rounded to two decimal places) is at least 0.45 m²K/W; or

b. for other edge designs:

R_{min} (rounded to two decimal places) is at least 0.20 m²K/W.

Note: These criteria do not apply to cases where internal surface projections are used to avoid surface condensation, eg curtain walling.

D7 In the event of an unacceptable risk being identified, marginal cases could be more rigorously anlaysed using numerical calculation methods, but in any case modification to improve the design should be considered.

Additional heat loss
D8 For the purposes of regulations for the conservation of fuel and power, the additional heat losses at the edges of openings may be ignored if:

a. for edges containing thin layers of thickness not exceeding 4 mm,

R_{mod} (rounded to two decimal places) is at least 0.45 m²K/W; or

b. for other edge designs:

R_{min} (rounded to two decimal places) is at least 0.45 m²K/W.

Compensating for additional heat loss
D9 Where the additional heat losses around the edges of openings cannot be ignored they can be taken into account in calculations as follows:

a. for dwellings the Target U-value method could be used with the average U-value increased by the following amount,

$$\frac{0.3 \times \text{total length of relevant opening surrounds}}{\text{total exposed surface area}} \text{ (W/m''K)}$$

b. for other buildings the calculation procedure could be used with the rate of heat loss from the

---

proposed building increased by the following amount:

0.3 x total length of relevant opening surrounds (W)

c. compensating measures, such as reducing the U-value of one of the elements of the construction, should then be incorporated so that:

i. for dwellings, the average U-value does not exceed the Target U-value; or

ii. for other buildings, the rate of heat loss from the proposed building does not exceed that of the notional building.

Example

Diagram D2 shows a window jamb in a masonry cavity wall with the blockwork returned towards the outer leaf at the reveal. By inspection it can be seen that ABC is the minimum resistance path.

Diagram D2 **70 mm cavity wall showing window jamb with blockwork returned at the reveal**

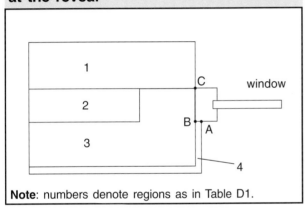

Note: numbers denote regions as in Table D1.

Table D1 **Thermal conductivity of materials in Diagram D2**

Region	Material	Conductivity (W/mK)
1	Brick outer leaf	0.84
2	Insulation	0.035
3	Medium weight concrete block inner leaf	0.61
4	Lightweight plaster	0.16

Calculation of R_{min}

Using the thermal conductivities from Table D1, Table D2 gives the resistance **R** for each segment of the path ABC. **R** for each segment is obtained by dividing the length of the path segment in metres by its thermal conductivity in W/mK. R_{min} is the sum of the resistances of each path segment.

Avoidance of the risk of surface condensation and mould growth

Referring to paragraph D6, R_{min} in this example is greater than 0.20 m²K/W and so the risk of surface condensation and mould growth is acceptably low.

Table D2 **Thermal resistance path in Diagram D2**

Path segments	Length (m)	Conductivity (W/mK)	R (m²K/W)
AB	0.015	0.16	0.094
BC	0.070	0.61	0.115
Minimum resistance R_{min}		=	0.209

Additional heat loss at the edge detail

Referring to paragraph D8, R_{min} in this example is less than 0.45 m²K/W, and so the additional heat loss at this edge should not be ignored.

Improving the edge design

Instead of returning the blockwork at the reveal the cavity could be closed using an insulated cavity closer, as in Diagram D3.

Diagram D3 **Window jamb showing cavity closed with an insulated cavity closer**

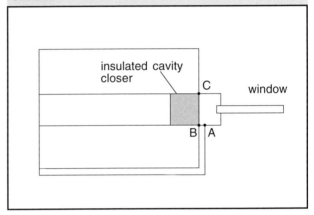

insulated cavity closer

The revised calculation of the minimum resistance is shown in Table D3. R_{min} is now greater than 0.45 m²K/W and so the additional heat loss can be ignored.

Table D3 **Minimum resistance path with insulated cavity closer**

Path segments	Length (m)	Conductivity (W/mK)	R (m²K/W)
AB	0.015	0.16	0.094
BC	0.070	0.04	1.750
Minimum resistance R_{min}		=	1.844

Alternative method

D10 A heat loss factor for a particular detail could be obtained by a numerical method and used to modify the calculation of the average U-value or the total rate of heat loss. A calculation procedure for deriving such loss factors is given in BRE IP 12/94, *Assessing condensation risk and heat loss at thermal bridges around openings.*

Appendix E

THE ELEMENTAL METHOD

Table E1 A way of demonstrating compliance of dwellings using the Elemental Method

What to do	How to use the Approved Document to check if your design is satisfactory according to the Elemental Method
1 Estimate if the SAP rating of the proposed dwelling is likely to be 60 or less.	The example dwellings in Appendix G will help you to estimate your SAP rating. If it is likely to be 60 or less, the Elemental Method gives more stringent U-values for roofs, floors and windows, doors and rooflights, as set out in Table 1 on page 8.
2 Check that the roof achieves a U-value of: (a) 0.25 (if the SAP rating is more than 60) (b) 0.2 (if the SAP rating is 60 or less).	Use the tables in Appendix A to calculate the thickness of insulation needed.[1]
3 Check that exposed walls achieve a U-value of 0.45 and the semi-exposed walls achieve a U-value of 0.6.	Use the tables in Appendix A to calculate the thickness of insulation needed.[1]
4 Check that the ground floor achieves a U-value of: (a) 0.45 (if the SAP rating is more than 60) (b) 0.35 (if the SAP rating is 60 or less).	(a) Calculate the perimeter to area ratio as explained in Appendix C (b) Use the perimeter/area ratio in the tables in Appendix A to calculate the thickness of insulation needed.[2]
5 Check that any exposed or semi-exposed upper floors achieve a U-value of: (a) 0.6 for semi-exposed floors (b) 0.45 for exposed floors (if the SAP rating is more than 60) (c) 0.35 for exposed floors (if the SAP rating is 60 or less).	Semi-exposed floors are defined in paragraph 0.13 on page 7. Use the tables in Appendix A to calculate the thickness of insulation needed.[1]
6 Calculate the average U-value of the windows, doors and rooflights.	Use Table 2 on page 8 to look up the U-values of your windows, doors and rooflights.[1] The example on the following page explains how to calculate the average U-value.
7 Check that the total area of windows, doors and rooflights does not exceed the permitted amount.	Using the average U-value from step 6 above, use Table 3 on page 9 to look up the maximum permitted area of windows, doors and rooflights. It is given as a percentage of the **dwelling floor area**.
8 Check that the design will achieve a SAP rating of more than 60 if you have made this assumption in steps 2, 4 and 5.	Complete the worksheet in Appendix G, or get a competent person to do this for you. (See paragraphs 0.19 and 0.20 on page 7.)

Notes:

1. Certified manufacturers' test data should be used where available.
2. Alternatively follow the procedures in Appendix C.

CALCULATIONS TO DETERMINE THE PERMISSIBLE AREA OF WINDOWS, DOORS AND ROOFLIGHTS IN THE ELEMENTAL METHOD

Example 1

E1   A proposed dwelling has a total floor area of 80 m². It has two half-single-glazed doors and double-glazed windows as described in Table E2.

Table E2   Window and door openings for a proposed dwelling

Element	Area (m²)	U-value (W/m²K)	Rate of heat loss per degree (W/K)
Windows	15.4	2.9	44.66
Two doors	3.8	3.7	14.06
Totals	19.2		58.72

E2   The Average U-value for windows and doors is given by the ratio:

$$\frac{\text{Total rate of heat loss per degree}}{\text{Total area of window and door openings}}$$

From Table E2 above this ratio is:

$$\frac{58.72}{19.2} = \textbf{3.1 W/m}^{\prime}\textbf{K}$$

E3   The total area of openings is:

$$\frac{19.2}{80} = 24\% \text{ of the } \textit{floor area}$$

E4   It can be seen from Table 3 on page 9 that the permitted area of windows, doors and rooflights for a U-value of 3.1 W/m²K is 21.5% of the *floor area* if the dwelling has a SAP rating of 60 or less, or 24% if the SAP rating is more than 60. The proposed design therefore satisfies the requirements of the Regulations only if the SAP rating is more than 60.

Example 2

E5   A proposed single-storey assembly building has plan dimensions of 16 m by 8 m. Its height to the eaves is to be 4 m and the open-trussed roof is to be double-pitched at 30° with insulation fixed between and over the rafters. Windows, personnel doors and rooflights are to be provided as indicated in Table E3 below.

Table E3 **Proposed assembly building: windows, personnel doors and rooflights**

Element	Area (m²)	U-value (W/m²K)	Rate of heat loss per degree (W/K)
Windows	60	3.6	216.0
Personnel doors	9.5	3.0	28.5
Totals for windows and personnel doors	69.5	—	244.5
Rooflights	25	3.8	—

E6   The average U-value for windows and doors is given by the ratio:

$$\frac{\text{Total rate of heat loss per degree}}{\text{Total area of windows plus personnel doors}} =$$

$$\frac{244.5}{69.5} = 3.5 \, \text{W/m}^2\text{K}$$

E7   The gross wall area of the proposed building is 192 m² and thus the area of windows and personnel doors is:

$$\frac{69.5}{192} = 36.2\% \text{ of the } \textit{wall area}$$

E8   The area of the roof in the plane of insulation is:

$$\frac{2 \times 4}{\text{Cos } 30°} \times 16 = 147.8 \, \text{m}^2$$

The rooflights are hence:

$$\frac{25}{147.8} = 16.9\% \text{ of the } \textit{roof area}$$

E9   The proposed design meets the requirements, therefore, because Table 8 on page 18 indicates that:

a.   the permitted area of windows and personnel doors having an average U-value of 3.5 W/m²K is 37% of the *wall area*.

b.   the permitted area of rooflights having a U-value of 3.8 W/m²K is 17% of the *roof area*.

Appendix F

THE TARGET U-VALUE METHOD

Table F1 **A way of demonstrating compliance for dwellings using the Target U-value Method**

What to do	How to use the Approved Document to check if your design complies with the Target U-value Method
1 Estimate if the SAP rating of the proposed dwelling is likely to be 60 or less.	The example dwellings in Appendix G will help you to estimate your SAP rating. If it is likely to be 60 or less, the Target U-value is lower.
2 Calculate the surface area and the U-value for each of the following parts of the dwelling: • roof • exposed walls • ground and other exposed floors • windows, doors and rooflights	Exclude semi-exposed elements from the calculations. (See paragraph 0.13 on page 7 for a definition of semi-exposed.) Tabulate the figures to determine: a) the total exposed surface area (m²) of the dwelling, and b) the total rate of heat loss per degree (W/K) for the dwelling. Table F2 on the following page illustrates the procedure.
3 Check that any semi-exposed elements achieve a U-value of 0.6 or less.	Semi-exposed elements should be insulated to at least the standard given in Table 1 on page 8.
4 Calculate the Average U-value.	Use the figures obtained in Step 2 to calculate the ratio: $$\frac{\text{Total rate of heat loss per degree}}{\text{Total external surface area}}$$ This is the Average U-value for the dwelling.
5 Calculate the Target U-value.	Use the total floor area of the dwelling and the total exposed surface area, from Step 2, to calculate the Target U-value as set out in paragraph 1.12 on page 10.
6 Compare the Average U-value for your design with the Target U-value.	If the Average U-value (from Step 4) does not exceed the Target U-value, (from Step 5), your design meets the requirements. If the Average U-value is more, the design should be modified by taking account of solar gains or providing a more efficient heating system as indicated in steps 7 and 8 and/or by altering the design of the fabric elements. An example of how to do this is set out on the following two pages.
7 If you wish to, adjust the Average U-value to take account of solar gains.	The window area from Step 2 can be reduced if the area of glazing facing south (±30°) exceeds that facing north (±30°). This is explained more fully in paragraph 1.16 on page 10.
8 If you wish to, adjust the Target U-value to take account of a more efficient heating system.	The Target U-value can be increased by up to 10% if, for example, a condensing boiler is used. Paragraphs 1.17 and 1.18 on page 10 explain this more fully.
9 Check that the design will achieve a SAP rating of more than 60 if you have made this assumption.	Complete the worksheet in Appendix G, or get a competent person to do this for you. (See paragraphs 0.19 and 0.20 on page 7.)

EXAMPLES ILLUSTRATING THE USE OF THE TARGET U-VALUE METHOD FOR DWELLINGS

Example 1 – A detached dwelling

F1 Consider the example in Diagram F1 which has details as given in Table F2. It is proposed to adopt the Target U-value method with U-values for the walls and roof a little higher (worse) than would otherwise be required in the **Elemental** method. The SAP Energy Rating is to be more than 60. The cavity walls are to have dry-linings and the windows and doors are to have metal frames with thermal breaks and sealed double-glazing with 12 mm air gaps.

Table F2 **Data for the detached dwelling**

Exposed element	Exposed surface area (m²)	U-value (W/m²K)	Rate of heat loss per degree (area x U-value) (W/K)
Floor	56.2	0.45	25.29
Windows	24.8	3.3	81.84
Doors	3.8	3.3	12.54
Walls	121.4	0.5	60.7
Roof	56.2	0.3	16.86
Totals	**262.4**		**197.23**

The Target
F2 From paragraph 1.12 on page 10 the Target U-value for dwellings with SAP Energy Ratings of more than 60 is given by:

$$\frac{\text{Total floor area x 0.64}}{\text{Total exposed surface area}} + 0.4$$

In this case, therefore, the target U-value is:

$$\frac{112.4 \text{ x } 0.64}{262.4} + 0.4 = \textbf{0.67 W/m}^2\textbf{K}$$

The Average U-value
F3 The average U-value for the dwelling is given by the ratio of the two values:

$$\frac{\text{Total rate of heat loss per degree}}{\text{Total external surface area}}$$

These values are calculated as shown in Table F2 above. For this example, therefore, the average U-value is:

$$\frac{197.23}{262.4} = \textbf{0.75 W/m}^2\textbf{K}$$

F4 The proposed design does not meet the requirements and modifications must be explored. Possibilities include improving the thermal resistance of the exposed walls and the windows and doors and taking account of the benefits of solar gain and more efficient space heating systems (as described in paragraphs 1.16 to 1.18 on page 10). For illustration purposes they are all considered as follows.

Diagram F1 **Plans of the detached dwelling**

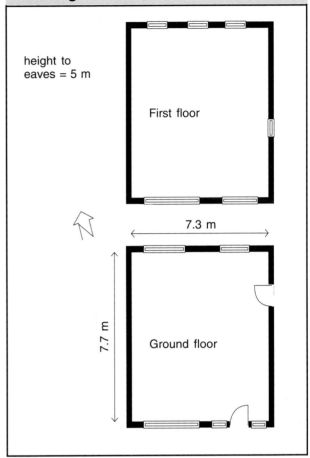

height to eaves = 5 m

First floor

7.3 m

7.7 m

Ground floor

Taking account of solar gain
F5 The total window area in the example is 24.8 m² of which 15.0 m² faces south ± 30° and 9.3 m² faces north ± 30°(the remaining 0.5 m² faces east). In accordance with paragraph 1.16 in the Approved Document therefore the area of window and hence the total area used in the calculation of the average U-value can be reduced by 40% of (15.0 – 9.3) = **2.28 m²**.

Improving the thermal resistance of the windows
F6 Table 2 on page 8 gives indicative figures for the U-values of various types of windows and doors although manufacturers' information should be used in preference if available. For the purposes of this example it is proposed to alter the window and door specifications to obtain U-values of 2.9 W/m²K and 3.0 W/m²K respectively.

Improving the thermal resistance of the walls
F7 Appendix A can be used to develop wall designs to achieve a range of U-values. For the purposes of this exercise it is proposed to alter the wall specification to obtain a U-value of 0.45 W/m²K.

Determination of the revised average U-value
F8 The revised average U-value is calculated in the same way as before. The revised table of data follows with the alterations highlighted.

Table F3 **Revised data for the detached dwelling**

Exposed element	Exposed surface area (m²)	U-value (W/m²K)	Rate of heat loss per degree (area x U-value) (W/K)
Floor	56.2	0.45	25.29
Windows	22.52	2.9	65.31
Doors	3.8	3.0	11.4
Walls	121.4	0.45	54.63
Roof	56.2	0.3	16.86
Totals	**260.12**		**173.49**

For the revised proposals, therefore, the average U-value is:

$$\frac{173.49}{260.12} = \textbf{0.67 W/m}^2\textbf{K}$$

Selecting a higher performance heating system

F9 In accordance with paragraph 1.17 on page 10 a hot water central heating system incorporating a condensing boiler could be specified in return for relaxing the target U-value by 10%. The target in this case would therefore be increased from 0.67 W/m²K to 0.74 W/m²K.

Compliance

F10 The detached dwelling as described in Diagram F1 and Table F2 could be made to comply with the requirements by, for example:

EITHER taking solar gain into account and increasing the performance of the windows, doors and walls,

OR increasing the performance of one of the elements, such as the windows, and installing a hot water central heating system incorporating a condensing boiler.

Example 2 – A semi-detached dwelling

F11 Consider the example in Diagram F2 which has details as given in Table F4. It is proposed to adopt the Target U-value approach with the walls having a U-value of 0.55 W/m²K. To compensate for this the windows and doors are to have an average U-value of 3.0 W/m²K. The SAP Energy Rating is to be more than 60.

F12 The party wall and the semi-exposed wall at the garage (which has a U-value of 0.6 W/m²K) are not included in the average U-value or Target U-value calculations.

The Target

F13 From paragraph 1.12 on page 10 the Target U-value for dwellings with SAP ratings of more than 60 is given by:

$$\frac{\text{Total floor area x 0.64}}{\text{Total exposed surface area}} + 0.4$$

In this case therefore the Target U-value is:

$$\frac{80 \times 0.64}{157} + 0.4 = \textbf{0.73 W/m}^{\prime}\textbf{K}$$

Diagram F2 **Plans of the semi-detached dwelling**

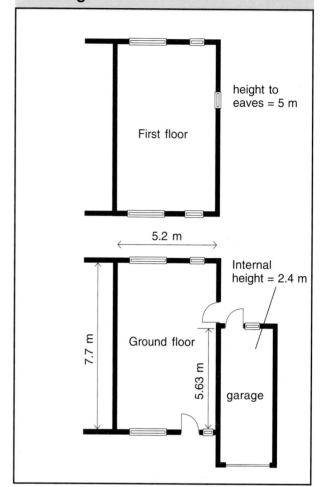

Table F4 **Data for the semi-detached dwelling**

Exposed element	Exposed surface area (m²)	U-value (W/m²K)	Rate of heat loss per degree (area x U-value) (W/K)
Floor	40.0	0.45	18.0
Windows	14.2	3.00	42.6
Doors	3.8	3.00	11.4
Walls	59.0	0.55	32.45
Roof	40.0	0.25	10.0
Totals	**157.0**		**114.45**

The Average U-value

F14 The average U-value for the dwelling is given by the ratio of the two values:

$$\frac{\text{Total rate of heat loss per degree}}{\text{Total external surface area}}$$

These values are calculated as shown in Table F3. For this example, therefore, the average U-value is:

$$\frac{114.45}{157.0} = \textbf{0.73 W/m}^{\prime}\textbf{K}$$

The proposed design therefore meets the Regulations.

Appendix G

THE SAP ENERGY RATING METHOD FOR DWELLINGS

The SAP Energy Rating and the Building Regulations

G1 This Appendix has been approved by the Secretary of State for the purpose of Regulation 14A of the Building Regulations 1991 as amended by the Building Regulations (Amendment) Regulations 1994.

G2 When calculating the SAP Energy Rating in compliance with Regulation 14A or for the purposes of following the guidance in Section 1 of Approved Document L the following considerations apply:

a. The data used in calculations of SAP Energy Ratings should be obtained from the tables in this Appendix. The fuel cost data will be revised in future editions of the Approved Document.

b. For each particular case the SAP Energy Rating should be calculated using the edition of this Appendix current at the date of giving notice or the deposit of plans in compliance with Regulation 11.

c. When the final heating system is unknown, the SAP Energy Rating notified to the building control body in accordance with Regulation 14A should be calculated assuming a main system of electric room heaters and a secondary system of electric heaters, both systems using on-peak electricity.

d. When undertaking SAP Energy Rating calculations for designs not intended for specific construction sites (eg type designs) the following assumptions should be made:

   i. two sides of the dwelling will be shaded;

   ii. the windows, doors and rooflights are all on the east and west elevations.

EXAMPLE SAP ENERGY RATINGS FOR DIFFERENT DWELLING TYPES

Example 1 – Two bedroom mid-terrace house

Table G1 **Data for the two bedroom mid-terrace house with electric storage heaters**

Element	Description	Area	U-value
Wall	Brick/cavity/dense block with 70 mm blown fibre cavity insulation	30.3	0.44
Roof	Pitched roof, 100 mm insulation between joists 50 mm on top	27.3	0.25
Ground floor	Suspended timber, 25 mm insulation	27.3	0.37
Windows and doors	Double glazed (6 mm gap), wooden frame	11.7	3.3
Heating	Electric storage heaters (efficiency 100%)		

SAP rating = 68

Diagram G1 **Plans of two bedroom mid-terrace house**

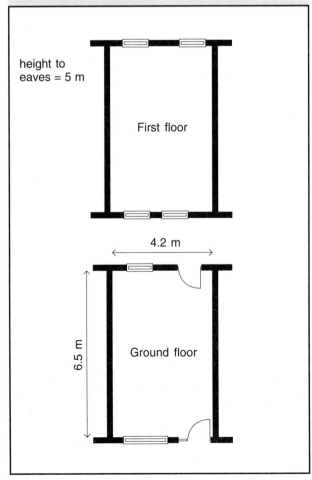

Example 2 – Three bedroom semi-detached house

Table G2 Data for the three bedroom semi-detached house with gas boiler

Element	Description	Area	U-value
Wall	Brick/cavity/dense block with 70 mm blown fibre cavity insulation	72.5	0.44
Roof	Pitched roof, 100 mm insulation between joists 50 mm on top	40	0.25
Ground floor	Solid concrete, 25 mm insulation	40	0.43
Windows and doors	Double glazed (6 mm gap), PVC-U frame	18.0	3.3
Heating	Central heating with gas boiler (efficiency 72%)		

SAP rating = 79

Example 3 – Three bedroom semi-detached house

Table G3 Data for the three bedroom semi-detached house with LPG boiler

Element	Description	Area	U-value
Wall	Brick/cavity/aerated concrete block with insulated plasterboard	72.5	0.45
Roof	Pitched roof, 100 mm insulation between joists 100 mm on top	40	0.19
Ground floor	Concrete suspended beam and aerated concrete block, 40 mm insulation	40	0.35
Windows and doors	Double glazed (12 mm gap), wooden frame	18.0	3.0
Heating	Central heating with LPG boiler (efficiency 72%)		

SAP rating = 58

Diagram G2 **Plans of three bedroom semi-detached house**

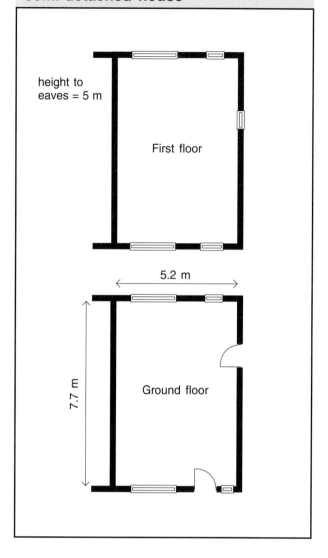

height to eaves = 5 m

First floor

5.2 m

7.7 m

Ground floor

Example 4 – Four bedroom detached house

Table G4 Data for the four bedroom detached house with condensing boiler

Element	Description	Area	U-value
Wall	Brick/partial cavity fill/ medium density block	116.5	0.45
Roof	Pitched roof, 100 mm insulation between joists 50 mm on top	50	0.25
Ground floor	Suspended timber, 35 mm insulation	50	0.45
Windows and doors	Double glazed (6 mm gap), wooden frame	24.9	3.3
Heating	Central heating with gas condensing boiler (efficiency 85%)		

SAP rating = 85

Diagram G3 Plans of four bedroom detached house

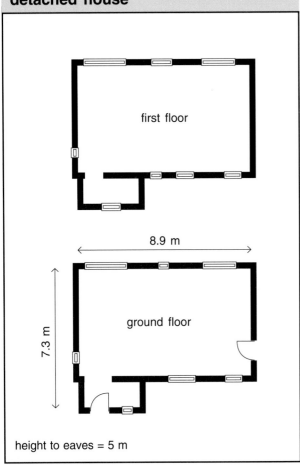

first floor

8.9 m

ground floor

7.3 m

height to eaves = 5 m

Example 5 – Two bedroom bungalow

Table G5 Data for the two bedroom bungalow with gas boiler

Element	Description	Area	U-value
Wall	Brick/cavity/aerated concrete block with insulated plasterboard	64.2	0.45
Roof	Pitched roof, 100 mm insulation between joists 50mm on top	56.7	0.25
Ground floor	Concrete suspended beam and medium density concrete block, 25 mm insulation	56.7	0.45
Windows and doors	Double glazed (6 mm gap), PVC-U frame	13.4	3.3
Heating	Central heating with gas boiler (efficiency 72%)		

SAP rating = 68

Diagram G4 Plan of two bedroom bungalow

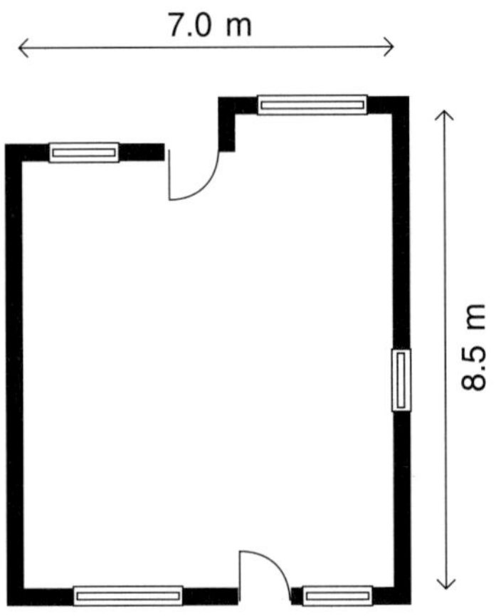

7.0 m

8.5 m

height to eaves = 2.4 m

THE STANDARD ASSESSMENT PROCEDURE

SUMMARY

This manual describes the Government's Standard Assessment Procedure (SAP) for producing an energy rating for a dwelling, based on calculated annual energy cost for space and water heating. The calculation assumes a standard occupancy pattern, derived from the measured floor area of the dwelling, and a standard heating pattern. The rating is normalised for floor area so that the size of the dwelling does not strongly affect the result, which is expressed on a scale of 1 – 100: the higher the number the better the standard.

The method of calculating the rating is set out in the form of a worksheet, accompanied by a series of tables. A calculation may be carried out by completing, in sequence, the numbered boxes in the worksheet, using the data in the tables as indicated. Alternatively, a computer program approved for SAP calculations by the Building Research Establishment may be used.

INTRODUCTION

An energy rating aims to inform householders of the overall energy efficiency of a home in a way that is simple and easy to understand. Many different technical definitions for such a rating are possible, and several have been used to date in the United Kingdom. The Standard Assessment Procedure (SAP) is the Government's recommended method for home energy rating.

The rating obtained from following the SAP depends upon a range of factors that contribute to energy efficiency:

- thermal insulation of the building fabric;
- efficiency and control of the heating system;
- ventilation characteristics of the dwelling;
- solar gain characteristics of the dwelling;
- the price of fuels used for space and water heating.

The rating is not affected by factors that depend on the individual characteristics of the household occupying the dwelling when the rating is calculated, for example:

- household size and composition;
- the ownership and efficiency of particular domestic electrical appliances;
- individual heating patterns and temperatures.

It is not affected either by the geographical location of the dwelling; a dwelling will have the same rating whether it is built in Caithness or Cornwall.

The procedure used to calculate the rating is based on the BRE Domestic Energy Model (BREDEM), which provides a framework for the calculation of energy use in dwellings. BREDEM exists in a number of standardised versions that differ in technical detail and in the precise requirements for data related to a particular application. Some versions are designed to be implemented as computer programs, while others are defined in such a way that they can be carried out without the use of a computer. Fuller details of BREDEM are given in earlier BRE reports.[1,2,3] The version described in this paper is a development of that in BRE Information Paper IP 13/88[4], and is suitable for calculating SAP ratings in both new and existing dwellings. When it is used for new dwellings to comply with Building Regulations, certain default values may be used in the absence of specific details of location, orientation, etc.

WORKSHEET CALCULATIONS

The method of calculating the rating is set out in the form of a worksheet, accompanied by a series of tables. A calculation is carried out by completing the numbered entries in the worksheet sequentially. Some entries are obtained by calculation from entries already made or by carrying forward an earlier entry; highlighting is used to indicate where that applies. Other entries are obtained, using linear interpolation where appropriate, by reference to Tables 1 to 14.

1. Dwelling dimensions

Dimensions refer to the inner surfaces of the elements bounding the dwelling. Thus floor dimensions are obtained by measuring between the inner surfaces of the external or party walls. This measurement should include all internal walls and built-in cupboards that are accessible from the occupied area of the dwelling. It should also include porches and conservatories where they are heated and form part of the habitable space. It should, however, exclude porches and conservatories where they are not heated and are clearly divided from the dwelling.

Storey height is the total height between the ceiling surface of a given storey and the ceiling surface of the storey below. For a single storey dwelling, or the ground floor of a dwelling with more than one storey, the measurement should be from floor surface to ceiling surface.

2. Ventilation rate

Chimneys and flues should be entered only when they are unrestricted and suitable for use. Balanced flues, such as those on many gas boilers and wall-mounted convector heaters, should not be included. Extract fans, including cooker hoods and other independent extractor fans should be included in the 'number of fans' category, but those that form part of a whole-dwelling mechanical ventilation system should be excluded.

Mechanical ventilation should only be entered if the whole dwelling is served by a mechanical ventilation system. Many whole-dwelling mechanical ventilation systems have a heat-exchanger to recover heat from the stale air being extracted from the dwelling. A warm-air heating system with flue heat recovery normally forms part of a full mechanical ventilation system with heat recovery.

In dwellings that have been subjected to a pressurisation test, steps (11) to (18) of the calculation may be by-passed. The air leakage at 50 Pa pressure difference, obtained from the test, is divided by 20 and the resulting value is entered in box (19).

When calculating for a new dwelling for Building Regulations purposes, enter 100% in box (16) for draught stripping of windows and doors. In the same context, it should be assumed that 2 sides of the dwelling are sheltered and hence the value 2 should be entered in box (20).

---

3. Heat losses

The areas of building elements are derived using the same principles as set out in section 1, being based on the internal dimensions of surfaces bounding the dwelling. Window area refers to the total area of the openings, including frames. Wall area is the net area of walls after allowing for windows and doors. Roof area is also net of any rooflights or windows set in the roof. The 'other' category is included principally to allow for the entry of heat losses to adjoining unheated areas, such as garages and conservatories. Box (35) can also be used to enter the sum of other heat losses calculated separately in those cases where there are several types of wall or roof. Losses or gains through party walls to spaces in other dwellings that are expected to be heated are assumed to be zero.

U-values for walls and roofs should be calculated using the proportional area method given in CIBSE Guide A3[5]. U-values for floors should be calculated using the procedure described in BRE Information Papers IP3/90[6] and IP7/93[7] and in Appendix C of the Building Regulations Approved Document L. U-values for windows should be obtained from Annex A unless certified manufacturer's data are available. This gives values appropriate for the whole window opening, including frames.

4. Water heating energy requirements

Demand for hot water is derived from the floor area of the dwelling and is given in Table 1. The energy required to produce that amount of hot water is then calculated, taking account of losses in generation, storage and distribution. The amount of heat released to the dwelling in the process of heating water is also estimated ('heat gains from water heating', box (52)) so that it can be taken into account in the calculation of space heating requirements.

A distinction is made between instantaneous water heating, which heats water when it is required, and water heating that relies on storage of hot water in a cylinder or tank. 'Primary' and 'cylinder' losses are not entered for instantaneous heaters. 'Single-point' heaters, which are located at the point of use and serve only one outlet, do not have distribution losses either. Gas 'multipoint' water heaters and 'combi' boilers are also instantaneous types but, as they normally serve several outlets, they are assumed to have distribution losses.

Stored hot water systems can either be served by an electric immersion heater or obtain heat from a boiler through a primary circuit. In both cases, cylinder losses are incurred to an extent that depends on how well the storage cylinder is insulated. Table 2 gives factors for calculating cylinder loss for different thicknesses of insulation. For boiler systems, primary losses are incurred in transferring heat from the boiler to the cylinder; values for primary losses are obtained from Table 3.

Water heating efficiency is obtained from Table 4. Note that the efficiency is reduced by 5% for systems that do not have a means of shutting down the boiler when the hot water cylinder reaches the required temperature. This function is normally carried out by a thermostat on the hot water cylinder that controls the boiler via a relay or the auxiliary contacts on a motorised control valve.

5. Internal gains

Internal gains from lights, appliances, cooking and from the occupants of the dwelling are estimated from floor area and obtained from Table 5. Gains from central heating pumps located within the heated space are added to the value obtained from Table 5, using the values given in the footnote to that table. Gains from the fans in a whole-dwelling mechanical ventilation system are also included but no useful gains are assumed from individual extractor fans.

6. Solar gains and utilisation factor

The solar gains entered are typical of the UK average and for dwellings where orientations are known should be obtained from Table 6. Gains for a particular location should not be used for this purpose, even though they might be considered to be more accurate for a particular case.

The solar access factor, box (65), is assumed to be 0.4 for a building that has heavy overshading, 0.7 more than average, 1.0 for average and 1.3 for a building with very little overshading.

For new dwellings being assessed for Building Regulations, where orientations have not been fixed, glazed elements should be entered for East/West orientations and the solar access factor set to 1.

The solar gains are added to the internal gains to give total heat gains. A utilisation factor is then applied to the gains, which has the effect of reducing the contribution of gains where they are large in relation to the heat load. This factor is calculated from the ratio of the total heat gains to the specific heat loss of the dwelling and is obtained from Table 7.

7. Mean internal temperature

The calculated mean internal temperature is based on the heating requirements of a typical household, taking account of the degree to which the dwelling is insulated and how well the heating can be controlled. The average temperature of the living area is obtained from Table 8, using the 'Heat loss parameter' (HLP), obtained from box (38) in the worksheet, and the type of heating system, obtained from the 'Heating' column of Table 4a. Note that the temperature obtained is raised in certain cases where the heating controls are poor; such cases are identified in Table 4b. This result is adjusted to take account of the level of heat gains previously calculated in sections 5 and 6 of the worksheet.

The temperature difference between the living area (zone 1) and the rest of the house is obtained from Table 9, using the HLP and the 'Control' column of Table 4b.

Zone 1 is defined as all rooms in the dwelling that are accessible from the living room without having to open any doors or to go up or down stairs. The fraction of the dwelling in Zone 1 may be measured or, where appropriate, can be estimated from the number of rooms in each zone, using the formula:

$$F = \frac{(1 + \text{number of rooms in Zone 1})}{(1 + \text{total number of rooms})}$$

For the purpose of this calculation, all rooms should be counted, but bathrooms and toilets should only count as half of one room each. The hall and landing count as a single room.

8. Degree-days

The degree-days depend on the 'base' temperature, which is calculated by adjusting the mean internal temperature to take account of the heat provided by gains. Degree-days for different base temperatures are obtained from Table 10, using linear interpolation for intermediate values.

9. Space heating requirements

The 'useful' energy required from the heating system is calculated from degree-days and specific heat loss. The quantity of fuel or electric energy required to meet that demand is then calculated, taking account of the efficiency of the space heating system (obtained from Table 4a). The procedure for dealing with dwellings that have more than one heating system is described in Annex B. Table 11 gives the combinations of heating systems that may be used in the SAP, together with the proportions of heat assumed to be supplied by main and secondary systems.

10. Fuel costs

Fuel costs are calculated using the fuel and electricity requirements calculated earlier in the worksheet and prices obtained from Table 12. The prices given in Table 12 are averaged over the previous three years and across regions. Other prices must not be used for the purpose of this calculation.

11. Energy cost rating

An energy cost deflator term is applied before the rating is calculated. The purpose of the deflator is to ensure that the ratings do not, on average, change with fuel price changes. The energy cost deflator will be updated periodically, as will the fuel prices in Table 12. The SAP rating is related to the energy cost factor by the equation:

$$SAP = 115 - 100 \times \log_{10}(ECF)$$

where: *ECF = Energy cost factor*

SAP ratings may also be obtained by using Table 14.

The SAP rating scale was chosen to cover a wide spectrum of energy costs, rising by one unit as cost falls by a constant percentage. A SAP rating of 1 represents a poor standard of energy efficiency while a rating of 100 represents a very high standard. With respect to the Building Regulations 1991 (amended 1994), a SAP rating of 60 or below indicates the need for a higher standard of fabric insulation. A SAP rating of more than the target specified in Table 4 in the Approved Document L is accepted as showing compliance with fabric insulation requirements, provided the guidance for limiting U-values and thermal bridging around openings is followed.

REFERENCES

1 **Anderson B R, A J Clarke, R Baldwin and N O Milbank,** *BREDEM The BRE Domestic Energy Model — background, philosophy and description.* BRE Report: BR 66, BRE, Watford, 1985.

2 **Henderson G and L D Shorrock,** *BREDEM - BRE Domestic Energy Model - testing the predictions of a two zone model,* Build. Serv. Eng. Res. Technol., 7(2) 1986, p87-91

3 **Shorrock L D, S Macmillan, J Clark and G Moore,** *BREDEM 8, a monthly calculation method for energy use in dwellings: testing and development,* Proceedings of the BEPAC conference on Building Environmental Performance, Canterbury, 1991

4 **Anderson B R,** *Energy assessment for dwellings using BREDEM worksheets,* BRE Information Paper IP13/88, BRE, Watford, 1988.

5 **CIBSE Guide A3,** The Chartered Institution of Building Services Engineers, London 1986.

6 **Anderson B R,** *The U-value of ground floors: application to Building Regulations,* BRE Information Paper IP3/90. Garston, BRE, 1990

7 **Anderson B R,** *The U-value of solid ground floors with edge insulation,* BRE Information Paper IP7/93. Garston, BRE, 1993.

SAP WORKSHEET (Version - 9.53)

1. Overall dwelling dimensions

	Area (m²)		Av. Room height(m)		Volume (m³)	
Ground floor	___	(1a) ×	___	=	___	(1b)
First floor	___	(2a) ×	___	=	___	(2b)
Second floor	___	(3a) ×	___	=	___	(3b)
Third and other floors	___	(4a) ×	___	=	___	(4b)

Total floor area $(1a) + (2a) + (3a) + (4a) =$ ___ (5)

Dwelling volume (m³) $(1b) + (2b) + (3b) + (4b) =$ ___ (6)

2. Ventilation rate

m³ per hour

Number of chimneys	___ × 40	= ___	(7)
Number of flues	___ × 20	= ___	(8)
Number of fans and passive vents	___ × 10	= ___	(9)

Infiltration due to chimneys, fans and flues

Air changes per hour

$$= ((7) + (8) + (9)) \div (6) = \text{___} \quad (10)$$

If a pressurisation test has been carried out, proceed to box (19)

Number of storeys ___ (11)

Additional infiltration $= ((11) - 1) \times 0.1 = \text{___}$ (12)

Structural infiltration 0.25 for timber frame
0.35 for masonry construction ___ (13)

If suspended wooden floor, enter 0.2 (unsealed)
0.1 (sealed) ___ (14)

If no draught lobby, enter 0.05 ___ (15)

Percentage of windows and doors draught stripped ___ (16)
Enter 100 for new dwellings which are to comply with Building Regulations

Window infiltration $= 0.25 - (0.2 \times (16) \div 100) = \text{___}$ (17)

Infiltration rate $= (10) + (12) + (13) + (14) + (15) + (17) = \text{___}$ (18)

If pressurisation test done, (measured $L_{50} \div 20$) + (10) $= \text{___}$ (19)

else let (19) = (18)

Number of sides on which sheltered ___ (20)
Enter 2 for new dwellings where location is not shown

Shelter factor $= 1 - 0.075 \times (20) = \text{___}$ (21)

If mechanical ventilation with heat recovery,

effective air change rate $= ((19) \times (21) + 0.17) = \text{___}$ (22)

(If no heat recovery, add 0.33 air changes per hour to value in box (22))

If natural ventilation, then air change rate $= (19) \times (21) = \text{___}$ (23)

If (23) $\geq$ 1, then (24) = (23) ___ (24)

else (24) $= 0.5 + ((23)^2 \times 0.5)$

Effective air change rate (enter (22) or (24)) ___ (25)

---

3. Heat losses and heat loss parameter

ELEMENT			Area (m²)		U-value (W/m²K)		A × U (W/K)	
Doors			⬚	×	⬚	=	⬚	(26)
Windows (single glazed)	0.9	×	⬚	×	⬚	=	⬚	(27)
Windows (double glazed)	0.9	×	⬚	×	⬚	=	⬚	(28)
Rooflights	0.9	×	⬚	×	⬚	=	⬚	(29)
Ground floor			⬚	×	⬚	=	⬚	(30)
Walls (type 1)			⬚	×	⬚	=	⬚	(31)
Walls (type 2)			⬚	×	⬚	=	⬚	(32)
Roof (type 1)			⬚	×	⬚	=	⬚	(33)
Roof (type 2)			⬚	×	⬚	=	⬚	(34)
Other			⬚	×	⬚	=	⬚	(35)

Ventilation heat loss	=	(25) × 0.33 × (6)	=	⬚	(36)
Specific heat loss	=	(26) + (27) + ... (35) + (36)	=	⬚	(37)
Heat loss parameter (HLP)	=	(37) ÷ (5)	=	⬚	(38)

4. Water heating energy requirements

GJ/year

Energy content of heated water (Table 1, column (a))   ⬚ (39)

   If instantaneous water heating at point of use, enter 0 in boxes (40), (43) and (48), and go to (49)

Distribution loss (Table 1, column (b))   ⬚ (40)

   If no hot water tank (ie combi or multipoint), go to (49)

Tank Volume   ⬚ (41)

Tank loss factor (Table 2)   ⬚ (42)

Energy lost from tank in GJ/year   (41) × (42) = ⬚ (43)

   If no solar panel, enter 0 in box (47) and go to (48)

Area of solar panel (m²)   ⬚ (44)

   Solar energy available =   1.3 × (44) ⬚ (45)

   Load ratio   =   (39) ÷ (45) = ⬚ (46)

   Solar input   = (45) × (46) ÷ (1 + (46)) = ⬚ (47)

Primary circuit loss (Table 3)   ⬚ (48)

Output from water heater   =   (39) + (40) + (43) + (48) − (47) = ⬚ (49)

Efficiency of water heater (Table 4) %   ⬚ (50)

Energy required for water heating   =   (49) ÷ (50) × 100 = ⬚ (51)

Heat gains from water heating   = 0.8 × ((40) + (43) + (48)) + 0.25 × (39) = ⬚ (52)

5. Internal gains

Watts

Lights, appliances, cooking and metabolic (Table 5) _____ (53)

Water heating = 31.7 × (52) = _____ (54)

Total internal gains = (53) + (54) = _____ (55)

6. Solar gains

Enter the area of the whole window including frames and the value for solar flux obtained from Table 6.
For Building Regulations assessment when orientations are not known, all vertical glazing should be entered as E/W.

Orientation	Area		Flux		Gains (W)	
North	_____	×	_____	=	_____	(56)
Northeast	_____	×	_____	=	_____	(57)
East	_____	×	_____	=	_____	(58)
Southeast	_____	×	_____	=	_____	(59)
South	_____	×	_____	=	_____	(60)
Southwest	_____	×	_____	=	_____	(61)
West	_____	×	_____	=	_____	(62)
Northwest	_____	×	_____	=	_____	(63)
Rooflights	_____	×	_____	=	_____	(64)

Solar access factor
Enter 1 for new dwellings where overshading is not known _____ (65)

Solar gains (UK average) = ((56) + (57) + ... + (64)) × (65) = _____ (66)

Total gains (W) (55) + (66) = _____ (67)

Gains/loss ratio (GLR) = (67) ÷ (37) = _____ (68)

Utilisation factor (Table 7) _____ (69)

Useful gains (W) = (67) × (69) = _____ (70)

7. Mean internal temperature

°C

Mean internal temperature of the living area (Table 8) _____ (71)

Adjustment for gains = 0.2 × R × ((70) ÷ (37) − 4.0) _____ (72)

   R is obtained from the 'responsiveness' column of Table 4a

Adjusted living room temperature = (71) + (72) = _____ (73)

Temperature difference between zones (Table 9) _____ (74)

Living area fraction (0 to 1.0) _____ (75)

Rest-of-house area fraction = 1.0 − (75) = _____ (76)

Mean internal temperature = (73) − ((74) × (76)) = _____ (77)

8. Degree-days

Temperature rise from gains $\qquad = \qquad$ (70) ÷ (37) = [ ] (78)

Base temperature $\qquad = \qquad$ (77) – (78) = [ ] (79)

Degree days (use (79) and Table 10 to adjust base) [ ] (80)

9. Space heating requirement **GJ/year**

Energy requirement (useful) $\qquad = \qquad$ 0.000 086 4 × (80) × (37) = [ ] (81)

Fraction of heat from secondary system [ ] (82)

Use value obtained from Table 11

Efficiency of main heating system (Table 4(a)) $\left.\begin{array}{l} \\ \\ \end{array}\right\}$ *Reduce by the amount shown in the 'efficiency' column of Table 4(b), where appropriate.* [ ] (83)

Efficiency of secondary heating system (Table 4(a)) [ ] (84)

Space heating fuel (main) $\qquad = \qquad$ (1.0 – (82)) × (81) × 100 ÷ (83) = [ ] (85)

Space heating fuel (secondary) $\qquad = \qquad$ (82) × (81) × 100 ÷ (84) = [ ] (86)

Electricity for pumps and fans
Enter 0.47 GJ for each central heating pump, 0.16 GJ for each boiler with a fan-assisted flue. For warm air heating system fans, add 0.002 GJ × the volume of the dwelling, (given in box (6)). For dwellings with whole-house mechanical ventilation, add 0.004 GJ × the volume of the dwelling. [ ] (87)

10. Fuel costs Fuel prices to be obtained from Table 12

			GJ/year ×	**Fuel price** =	**£/year**	
Space heating	- main system	=	(85) ×	[ ] =	[ ]	(88)
	- secondary system	=	(86) ×	[ ] =	[ ]	(89)

Water heating

If off-peak electric water heating: **Fuel price** **£/year**

On-peak percentage (Table 13) [ ] (90)

Off-peak percentage 100 – (90) = [ ] (91)

On-peak cost $\qquad = \qquad$ (51) × ((90) ÷ 100) × [ ] = [ ] (92)

Off-peak cost $\qquad = \qquad$ (51) × ((91) ÷ 100) × [ ] = [ ] (93)

Else:

Water heating cost $\qquad = \qquad$ (51) × [ ] = [ ] (94)

Pump/fan energy cost $\qquad = \qquad$ (87) × [ ] = [ ] (95)

Additional standing charges (Table 12) [ ] (96)

Total heating $\qquad$ (88) + (89) + (92) + (93) + (94) + (95) + (96) = [ ] (97)

11. SAP rating

Energy cost deflator (Table 12 footnote[2]) [ ] (98)

Energy cost factor (ECF) $\qquad = \qquad$ ((97) × (98) – 40.0) ÷ (5) = [ ] (99)

SAP rating (Table 14) [ ]

TABLES

Table 1 : Hot water energy requirements

Floor area (m²)	(a) Hot water usage GJ/year	(b) Distribution loss GJ/year
30	4.13	0.73
40	4.66	0.82
50	5.16	0.91
60	5.68	1.00
70	6.17	1.09
80	6.65	1.17
90	7.11	1.26
100	7.57	1.34
110	8.01	1.41
120	8.44	1.49
130	8.86	1.56
140	9.26	1.63
150	9.65	1.70
160	10.03	1.77
170	10.40	1.84
180	10.75	1.90
190	11.10	1.96
200	11.43	2.02
210	11.74	2.07
220	12.05	2.13
230	12.34	2.18
240	12.62	2.23
250	12.89	2.27
260	13.15	2.32
270	13.39	2.36
280	13.62	2.40
290	13.84	2.44
300	14.04	2.48

Alternatively, requirements and gains may be calculated from the total floor area of the dwelling (TFA), using the following steps:

(a) Calculate $N = 0.035 \times TFA - 0.000\,038 \times TFA^2$, if TFA ≤ 420
$\qquad = 8.0$ if TFA > 420

(b) Hot water usage $= 0.85 \times (92 + 61 \times N) \div 31.71$

(c) Distribution loss $= 0.15 \times (92 + 61 \times N) \div 31.71$

Table 2 : Hot water cylinder loss factor (GJ/year/litre)
Multiply by cylinder volume in litres to get loss

Insulation thickness mm	Type	
	Foam	Jacket
None	0.0945	0.0945
12.5	0.0315	0.0725
25	0.0157	0.0504
38	0.0104	0.0331
50	0.0078	0.0252
80	0.0049	0.0157
100	0.0039	0.0126
150	0.0026	0.0084

Table 3 : Primary circuit losses (GJ/year)

System loss	GJ/year
Electric immersion heater	0.0
Boiler with uninsulated primary pipework and no cylinder stat*	4.4
Boiler with insulated primary pipework and no cylinder stat*	2.2
Boiler with uninsulated primary pipework and with cylinder stat	2.2
Boiler with insulated primary pipework and with cylinder stat	1.3

A cylinder stat is required by the Building Regulations in new dwellings.

Table 4a: Heating system efficiency

This table shows space heating efficiency. Water heating efficiency is also obtained from this table when hot water is from a boiler system. For independent electric water heating use 100%. Use 70% for single-point gas water heaters. For multi-point gas water heaters and for heat exchangers built into gas warm air heaters, use 65%.

	Efficiency (%)	Heating type	Responsive-ness
CENTRAL HEATING SYSTEMS WITH RADIATORS			

1. 'Heating type' refers to the appropriate column in Table 8
2. Refer to Group 1 in Table 4b for control options
3. Check Table 4b for efficiency adjustment due to poor controls
4. Where two figures are given, the first is for space heating and the second is for water heating

	Efficiency (%)	Heating type	Responsive-ness
Gas boilers (including LPG) with fan-assisted flue			
1. Low thermal capacity	72	1	1.0
2. High or unknown thermal capacity	68	1	1.0
3. Condensing	85	1	1.0
4. Combi	71/69	1	1.0
5. Condensing combi	85/83	1	1.0
Gas boilers (including LPG) with balanced or open flue			
1. Wall mounted	65	1	1.0
2. Floor mounted, post 1979	65	1	1.0
3. Floor mounted, pre 1979	55	1	1.0
4. Combi	65	1	1.0
5. Room heater + back boiler	65	1	1.0
Oil boilers			
Standard oil boiler pre-1985	65	1	1.0
Standard oil boiler 1985 or later	70	1	1.0
Condensing boiler	85	1	1.0
Solid fuel boilers			
Manual feed (in heated space)	60	2	0.75
Manual feed (in unheated space)	55	2	0.75
Autofeed (in heated space)	65	2	0.75
Autofeed (in unheated space)	60	2	0.75
Open fire with back boiler to rads	55	3	0.50
Closed fire with back boiler to rads	65	3	0.50
Electric boilers			
Dry core boiler in heated space	100	2	0.75
Dry core boiler in unheated space	85	2	0.75
Economy 7 boiler in heated space	100	2	0.75
Economy 7 boiler in unheated space	85	2	0.75
On-peak heat pump	250	1	1.0
24 hour heat pump	240	1	1.0
STORAGE RADIATOR SYSTEMS			

(Refer to Group 2 in Table 4b for control options)

	Efficiency (%)	Heating type	Responsive-ness
Old (large volume) storage heaters	100	5	0.0
Modern (slimline) storage heaters	100	4	0.25
Convector storage heaters	100	4	0.25
Fan-assisted storage heaters	100	3	0.5
Electric underfloor heating	100	5	0.0

	Efficiency (%)	Heating type	Responsive-ness
WARM AIR SYSTEMS			

1. Refer to Group 3 in Table 4b for control options

	Efficiency (%)	Heating type	Responsive-ness
Gas-fired warm air with fan-assisted flue			
Ducted, with gas-air modulation	80	1	1.0
Room heater, with in-floor ducts	77	1	1.0
Gas-fired warm air with balanced or open flue			
Ducted (on/off control)	70	1	1.0
Ducted (modulating control)	72	1	1.0
Stub ducted	70	1	1.0
Ducted with flue heat recovery	85	1	1.0
Stub ducted with flue heat recovery	82	1	1.0
Condensing	94	1	1.0
Oil-fired warm air			
Ducted output (on/off control)	70	1	1.0
Ducted output (modulating control)	72	1	1.0
Stub duct system	70	1	1.0
Electric warm air			
Electricaire system	100	2	0.75
Air-to-air heat pump	250	1	1.0
ROOM HEATER SYSTEMS			

1. Refer to Group 4 in Table 4b for control options
2. Check Table 4b for efficiency adjustment due to poor control

	Efficiency (%)	Heating type	Responsive-ness
Gas			
Old style gas fire (open front)	50	1	1.0
Modern gas fire (glass enclosed front)	60	1	1.0
Modern gas fire with balanced flue	70	1	1.0
Modern gas fire with back boiler (no rads)	65	1	1.0
Condensing gas fire (fan-assisted flue)	85	1	1.0
Gas fire or room heater with fan-assisted flue	79	1	1.0
Coal effect fire in fireplace	25	1	1.0
Coal effect fire in front of fireplace	60	1	1.0
Solid fuel			
Open fire in grate	32	3	0.5
Open fire in grate, with throat restrictor	42	3	0.5
Open fire with back boiler (no rads)	55	3	0.5
Closed room heater	60	3	0.5
Closed room heater with back boiler	65	3	0.5
Electric (direct acting)			
Panel, convector or radiant heaters	100	1	1.0
Portable electric heaters	100	1	1.0
OTHER SYSTEMS			

Refer to group 5 for control options

	Efficiency (%)	Heating type	Responsive-ness
Gas underfloor heating	70	4	0.25
Gas underfloor heating, condensing boiler	87	4	0.25
Electric ceiling heating	100	2	0.75

Table 4b: Heating system controls

1. Use Table 4a to select appropriate Group in this table.

2. 'Control' refers to the appropriate column in Table 9.

3. 'Efficiency' is an adjustment that should be subtracted from the space heating efficiency obtained from Table 4a.

4. 'Temp' is an adjustment that should be added to the temperature obtained from Table 8.

Type of control	Control	Efficiency %	Temp °C
GROUP 1 : BOILER SYSTEMS WITH RADIATORS			
No thermostatic control* of room temperature	1	5	0.3
Programmer + roomstat	1	0	0
Prog + roomstat* (no boiler off)	1	5	0
Prog + roomstat + TRVs	2	0	0
Prog + roomstat + TRVs* (no boiler off)	2	5	0
TRVs + prog + bypass*	2	5	0
TRVs + prog + flow switch	2	0	0
TRVs + prog + boiler energy manager	2	0	0
Zone control	3	0	0
Zone control (no boiler off)*	3	5	0
GROUP 2 : STORAGE RADIATOR SYSTEMS			
Manual charge control	3	0	0.3
Automatic charge control	3	0	0
GROUP 3 : WARM AIR SYSTEMS			
No thermostatic control* of room temperature	1	0	0.3
Roomstat only*	1	0	0
Programmer + roomstat	1	0	0
Zone control	3	0	0
GROUP 4 : ROOM HEATER SYSTEMS			
No thermostatic control	2	0	0.3
Appliance stat	3	0	0
Appliance stat + prog	3	0	0
Programmer + roomstat	3	0	0
Roomstat only	3	0	0
GROUP 5 : OTHER SYSTEMS			
No thermostatic control of room temperature	1	0	0.3
Appliance stat only	1	0	0
Roomstat only	1	0	0
Programmer + roomstat	1	0	0
Programmer and zone control	3	0	0

* These systems would not be acceptable in new dwellings

Table 5 : Lighting, appliances, cooking and metabolic gains

Floor area (m²)	Gains (W)	Floor area (m²)	Gains (W)
30	230	170	893
40	282	180	935
50	332	190	978
60	382	200	1020
70	431	210	1061
80	480	220	1102
90	528	230	1142
100	576	240	1181
110	623	250	1220
120	669	260	1259
130	715	270	1297
140	760	280	1334
150	805	290	1349
160	849	300	1358

Alternatively, gains may be calculated from the total floor area of the dwelling (TFA), using the following steps:

(a) Calculate N $= 0.035 \times \text{TFA} - 0.000\,038 \times \text{TFA}^2$, if $\text{TFA} \le 420$
$= 8.0$ if $\text{TFA} > 420$

(b) Gains (W) $= 74 + 2.66 \times \text{TFA} + 75.5 \times N$, if $\text{TFA} \le 282$
$= 824 + 75.5 \times N$ if $\text{TFA} > 282$

Note:
Gains from the following equipment should be added to the totals given above:

Central heating pump	10 W
Mechanical ventilation system	25 W

Table 6 : Solar flux through glazing (W/m²)

	Horizontal		Vertical			
		North	NE/NW	E/W	SE/SW	South
Single glazed	31	10	14	18	24	30
Double glazed	26	8	12	15	21	26
Double glazed with low-E coating	24	8	11	14	19	24
Triple glazed	22	7	10	13	17	22

Note:
1. For a rooflight in a pitched roof with a pitch of up to 70°, use the value under 'Northerly' for orientations within 30° of North and the value under 'Horizontal' for all other orientations. (If the pitch is greater than 70°, then treat as if it were a vertical window.)

2. For Building Regulations assessment when orientations are not known, all vertical glazing should be entered as E/W.

Table 7 : Utilisation factor as a function of Gain/loss ratio (G/L)

G/L	Utilisation factor	G/L	Utilisation factor
1	1.00	16	0.68
2	1.00	17	0.65
3	1.00	18	0.63
4	0.99	19	0.61
5	0.97	20	0.59
6	0.95	21	0.58
7	0.92	22	0.56
8	0.89	23	0.54
9	0.86	24	0.53
10	0.83	25	0.51
11	0.81	30	0.45
12	0.78	35	0.40
13	0.75	40	0.36
14	0.72	45	0.33
15	0.70	50	0.30

Alternatively, the utilisation factor may be calculated by the formula:

$$\text{Utilisation factor} = 1 - \exp(-18 \div GLR),$$

where GLR = (total gains) ÷ (specific heat loss)

Table 8 : Mean internal temperature of living area

Number in brackets is from the 'heating' column of Table 4a. HLP is item 38 on the worksheet

HLP	(1)	(2)	(3)	(4)	(5)
1.0 (or lower)	18.88	19.32	19.76	20.21	20.66
1.5	18.88	19.31	19.76	20.20	20.64
2.0	18.85	19.30	19.75	20.19	20.63
2.5	18.81	19.26	19.71	20.17	20.61
3.0	18.74	19.19	19.66	20.13	20.59
3.5	18.62	19.10	19.59	20.08	20.57
4.0	18.48	18.99	19.51	20.03	20.54
4.5	18.33	18.86	19.42	19.97	20.51
5.0	18.16	18.73	19.32	19.90	20.48
5.5	17.98	18.59	19.21	19.82	20.45
6.0 (or higher)	17.78	18.44	19.08	19.73	20.40

Table 9 : Difference in temperatures between zones

Number in brackets is from the 'control' column of Table 4b. HLP is item 38 on the worksheet

HLP	(1)	(2)	(3)
1.0 (or lower)	0.40	1.41	1.75
1.5	0.60	1.49	1.92
2.0	0.79	1.57	2.08
2.5	0.97	1.65	2.22
3.0	1.15	1.72	2.35
3.5	1.32	1.79	2.48
4.0	1.48	1.85	2.61
4.5	1.63	1.90	2.72
5.0	1.76	1.94	2.83
5.5	1.89	1.97	2.92
6.0 (or higher)	2.00	2.00	3.00

Table 10 : Degree-days as a function of base temperature

Base temperature °C	Degree-days	Base temperature °C	Degree-days
1.0	0	11.0	1140
1.5	30	11.5	1240
2.0	60	12.0	1345
2.5	95	12.5	1450
3.0	125	13.0	1560
3.5	150	13.5	1670
4.0	185	14.0	1780
4.5	220	14.5	1900
5.0	265	15.0	2015
5.5	310	15.5	2130
6.0	360	16.0	2250
6.5	420	16.5	2370
7.0	480	17.0	2490
7.5	550	17.5	2610
8.0	620	18.0	2730
8.5	695	18.5	2850
9.0	775	19.0	2970
9.5	860	19.5	3090
10.0	950	20.0	3210
10.5	1045	20.5	3330

Table 11 : Fraction of heat supplied by secondary heating systems

Main heating system	Secondary system	Fraction
Central heating system with boiler and radiators, central warm-air system or other gas fired systems	gas fires	0.15
	coal fires	0.10
	electric heaters	0.05
Gas room heaters	gas fires	0.30
	coal fires	0.15
	electric heaters	0.10
Coal room heaters or electric room heaters	gas fires	0.20
	coal fires	0.20
	electric heaters	0.20
Electric storage heaters or other electric systems	gas fires	0.15
	coal fires	0.10
	electric heaters	0.10
Electric heat pump systems with heat storage or fan-assisted storage heaters	gas fires	0.15
	coal fires	0.10
	electric heaters	0.05

Table 12 : Fuel prices and additional standing charges

	Additional standing charge £	Unit price £/GJ
Gas (mains)	38	4.26
Bulk LPG	51	7.11
Bottled gas – propane 47 kg cylinder		13.26
Heating oil		3.68
House coal		3.87
Smokeless fuel		6.54
Anthracite nuts		5.04
Anthracite grains		4.53
Wood		4.20
Electricity (on-peak)		22.38
Electricity (off-peak)	14	7.60
Electricity (standard tariff)		21.08
Electricity (24-hr heating tariff)	45	8.57

Notes

1. The standing charge given for electricity is extra amount for the off-peak tariff, over and above the amount for the standard domestic tariff, as it is assumed that the dwelling has a supply of electricity for reasons other than space and water heating. Standing charges for gas and for off-peak electricity are attributed to space and water heating costs where those fuels are used for heating.

2. The energy cost deflator term is currently set at 0.96. It will vary with the weighted average price of heating fuels in future, in such a way as to ensure that the SAP is not affected by the general rate of inflation. However, individual SAP ratings are affected by relative changes in the price of particular heating fuels.

Table 13 : On-peak fraction for electric water heating

Dwelling floor area m²	Cylinder size (litres) 110	160	210
40 or less	12 (56)	6 (18)	1
60	14 (58)	7 (21)	3
80	17 (60)	9 (24)	4
100	19 (62)	10 (27)	5
120	21 (63)	12 (30)	6
140	23 (65)	13 (33)	6
160	25 (66)	15 (35)	7
180	27 (68)	16 (37)	8
200	29 (69)	17 (40)	9
220	30 (70)	18 (42)	9
240	32 (71)	19 (44)	10
260	33 (72)	20 (45)	11
280	34 (73)	21 (43)	11
300	36 (74)	21 (45)	12
320	36 (75)	22 (46)	12
340	37 (75)	23 (47)	12
360	38 (76)	23 (48)	13
380	39 (76)	24 (49)	13
400	39 (76)	24 (49)	13
420 or more	39 (77)	24 (50)	13

Table 13 shows percentage of electricity required at on-peak rates for cylinders with dual immersion heaters. The figures in brackets are for cylinders with single immersion heaters.

Table 14 : SAP rating by energy cost factor

ECF £/m²	SAP rating	ECF £/m²	SAP rating
1.4	100	4	55
1.5	97	4.5	50
1.6	95	5	45
1.7	92	5.5	41
1.8	89	6	37
1.9	87	6.5	34
2	85	7	30
2.2	81	7.5	27
2.4	77	8	25
2.6	74	8.5	22
2.8	70	9	20
3	67	10	15
3.3	63	11	11
3.6	59	12	7
3.9	54	13	4
		14	1

The values in the above table may be obtained by using the formula:

$$\text{SAP Rating} = 115 - 100 \times \log_{10}(\text{ECF})$$

ANNEX A : U-VALUES FOR GLAZING

Table A1: Indicative U-values [W/m²K] for windows, doors and rooflights

| | Type of frame | | | | | | | |
| | Wood | | Metal | | Thermal break | | PVC-U | |
Air gap in sealed unit (mm)	6	12	6	12	6	12	6	12
Window, double-glazed	3.3	3.0	4.2	3.8	3.6	3.3	3.3	3.0
Window, double-glazed, low-E	2.9	2.4	3.7	3.2	3.1	2.6	2.9	2.4
Window, double-glazed, Argon fill	3.1	2.9	4.0	3.7	3.4	3.2	3.1	2.9
Window, double-glazed, low-E, Argon fill	2.6	2.2	3.4	2.9	2.8	2.4	2.6	2.2
Window, triple-glazed	2.6	2.4	3.4	3.2	2.9	2.6	2.6	2.4
Door, half double-glazed	3.1	3.0	3.6	3.4	3.3	3.2	3.1	3.0
Door, fully double-glazed	3.3	3.0	4.2	3.8	3.6	3.3	3.3	3.0
Rooflights, double-glazed at less than 70° from horizontal	3.6	3.4	4.6	4.4	4.0	3.8	3.6	3.4
Windows and doors, single-glazed	4.7		5.8		5.3		4.7	
Door, solid timber panel or similar	3.0		—		—		—	
Door, half single-glazed/half timber panel or similar	3.7		—		—		—	

Window U-values should be obtained from Table A1, regardless of building age. The values apply to the entire area of the window opening, including both frame and glass, and take account of the proportion of the area occupied by the frame and the heat conducted through it.

ANNEX B : DWELLINGS WITH MORE THAN ONE HEATING SYSTEM

General principles

The treatment of multiple forms of heating in SAP is based on the following principles:

(1) the decision to include a secondary heating system should be based on the characteristics of the dwelling and the systems installed and not on the heating practices of the occupying household;

(2) secondary systems should only be included if they are based on 'fixed' appliances, unless portable appliances are necessary to achieve adequate heating.

To avoid excessive complexity and to reduce the extent to which surveyor judgement can influence the rating, the SAP considers only one secondary heating system per dwelling. Furthermore, secondary heating systems are governed by a set of rules that restrict the allowable combinations of heating systems and stipulate the proportion of heat supplied by the secondary system. Those restrictions will inevitably mean that the rating is based on assumptions about use that, in some cases, diverge considerably from what is actually practised by the occupying household. This is in line with the general principle that SAP is a rating for the dwelling and does not depend on who happens to be living in that dwelling. That does not preclude further estimates of energy consumption being made to take account of actual usage. Such estimates would not be part of SAP but could form the basis of advice given to the occupying household on how to make best use of the systems at their disposal.

Procedure for dealing with secondary heating systems

(1) Identify the main heating system. If there is a central system that provides both space and water heating and it is capable of heating at least 30% of the dwelling, select that system. If there is no system that provides both space and water heating, then select the system that has the capability of heating the greatest part of the dwelling.

(2) If there is still doubt about which system should be selected as the primary, select the system that supplies useful heat to the dwelling at lowest cost, (obtained by dividing fuel cost by conversion efficiency).

(3) Use the responsiveness of the main heating system in calculating the mean internal temperature in Stage 7 of the SAP calculation.

(4) Decide whether a secondary heating system needs to be specified, bearing in mind that systems based on stored heat produced from electricity generally require a secondary system for their successful operation. Ignore all portable heaters, such as plug-in electrical radiators and fan heaters or free-standing butane and paraffin heaters. If no secondary system is to be specified, enter zero in box (82).

(5) If a secondary heating system is to be specified, use Table 11 to select the most appropriate description of the primary/secondary combination. Obtain the proportion of use for the secondary heating system from Table 11 and enter in box (82).

(6) Obtain the efficiency of the secondary heating system from Table 4 and enter in box (84).

(7) Calculate the space heating fuel requirements for both main and secondary heating systems as specified for entry in boxes (85) and (86).

Dwellings with inadequate heating systems

The SAP assumes that a good standard of heating will be achieved throughout the dwelling. For dwellings in which the heating system is not capable of providing that standard, it should be assumed that the additional heating is provided by electric heaters, using the fractions given in Table 11. For new dwellings that have no heating system specified, it should be assumed that all heat will be provided by electric heaters using electricity at the standard domestic tariff.

Appendix H

EXAMPLE CALCULATIONS FOR
BUILDINGS OTHER THAN
DWELLINGS

Example 1 – Calculation procedure for showing the fabric insulation meets the requirements

A detached, four storey office building 45 m x 13 m in plan and height 15 m is to be constructed with 55% glazing, using windows which have metal frames with thermal breaks containing sealed double-glazed units with 12 mm air gaps and a low-E coating. No rooflight glazing is proposed. The remaining exposed walls and the roof are to have U-values of 0.6 W/m²K and 0.45 W/m²K respectively, with the ground floor being uninsulated.

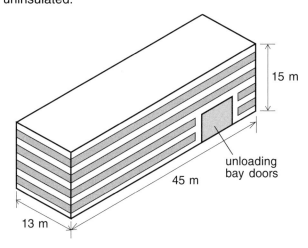

Proposed building

1 Calculate the areas of each element:

area of roof (45 x 13)	= 585 m²
area of elevations (45 + 45 + 13 + 13) x 15	= 1740 m²
area of windows (55% of 1740 m²)	= 957 m²
area of personnel doors	= 14 m²
area of vehicle unloading bay doors	= 27 m²
area of exposed wall (1740 – 957 – 14 – 27)	= 742 m²
area of floor (45 x 13)	= 585 m²

2 The rate of heat loss from the proposed building is calculated as follows:

Element	Area (m²)	U-value (W/m²K)	Rate of heat loss (W/K)
Roof	585	0.45	263.3
Exposed walls	742	0.6	445.2
Windows	957	2.6	2488.2
Personnel doors	14	3.3	46.2
Vehicle loading bay doors	27	0.7	18.9
Ground floor	585	0.36	210.6
Total rate of heat loss			3472.4

Notional building

The area of openings in the proposed building is more than the basic allowance in Table 5. So the basic area allowance of 40% should be assumed for the notional building. The notional building can, however, include the allowance for rooflights.

The exposed perimeter of the ground floor of the proposed building is 116 m and its area is 585 m². So the ratio P/A is 0.2, which indicates the U-value of the ground floor is 0.36 W/m²K (from Table C1 on page 40). This is less than the standard U-value in Table 5 so 0.36 W/m²K must be the value used in the notional building calculation.

1 Calculate the areas of each element:

area of rooflights (20% of 585 m²)	= 117 m²
area of roof (45 x 13) – 117	= 468 m²
area of elevations (45 + 45 + 13 + 13) x 15	= 1740 m²
area of windows and personnel doors (40% of 1740 m²)	= 696 m²
area of vehicle unloading bay doors	= 27 m²
area of exposed wall (1740 – 696 – 27)	= 1017 m²
area of floor (45 x 13)	= 585 m²

2 The rate of heat loss from the notional building is calculated as follows:

Element	Area (m²)	U-value (W/m²K)	Rate of heat loss (W/K)
Rooflights	117	3.3	386.1
Roof	468	0.45	210.6
Exposed walls	1017	0.45	457.7
Windows and personnel doors	696	3.3	2296.8
Vehicle loading bay doors	27	0.7	18.9
Ground floor	585	0.36	210.6
Total rate of heat loss			3580.7

The rate of heat loss from the proposed building is less than that from the notional building and therefore the requirements of the Regulations are satisfied.

Example 2 – Lighting calculation procedure to show that 95% of lamps are listed in Table 9

A new hall and changing rooms are to be added to an existing community centre. The proposed lighting scheme incorporates lamps that are listed in Table 9 except for some low voltage tungsten halogen downlighters which are to be installed in the entrance area with local controls. A check therefore has to be made to show that the low voltage tungsten halogen lamps comprise less than 5% of the overall installed capacity of the lighting installation.

Main hall
Twenty wall mounted uplighters with 250 W high pressure Sodium lamps are to provide general lighting needs. The uplighters are to be mounted 7 m above the floor. On plan, the furthest light is 20.5 m from its switch which is less than three times the height of the light above the floor.

It is also proposed to provide twenty 18 W compact fluorescent lights as an additional system enabling instant background lighting whenever needed.

Changing rooms, corridors and entrance
Ten 58 W, high frequency fluorescent light fittings are to be provided in the changing rooms and controlled by occupancy detectors. Six more 58 W fluorescent light fittings are to be located in the corridors and the entrance areas and switched locally. Additionally, in the entrance area there are to be the six 50 W tungsten halogen downlighters noted above.

Calculation
A schedule of light fittings is prepared as follows:

Position	Number	Description of light source	Circuit Watts per lamp	Total circuit Watts (W)
Main hall	20	250 W SON	286 W	5720
Main hall	20	18 W compact fluorescent	23 W	460
Entrance, changing rooms and corridors	16	58 W HF fluorescent	64 W	1024
Entrance	6	50 W low voltage tungsten halogen	55 W	330
			Total =	7534 W

The percentage of circuit Watts consumed by lamps not listed in Table 9:

$$= \frac{330 \times 100}{7534} = 4.4\%$$

Therefore, more than 95% of the installed lighting capacity, in circuit Watts, is from light sources listed in Table 9. The switching arrangements comply with paragraph 2.47. The proposed lighting scheme therefore meets the requirements of the Regulations.

Example 3 – Lighting calculation procedure to show average circuit efficacy is not less than 50 lumens/Watt

A new lighting scheme is proposed for a restaurant comprising a mixture of concealed perimeter lighting using high frequency fluorescent fittings and individual tungsten lamps over tables. Lights in the dining area are to be switched locally from behind the bar. The over-table lamps also have integral switches for diners' use. Lighting to kitchens and toilets is to be switched locally.

The table opposite shows a schedule of the light sources proposed together with a calculation of the overall average circuit efficacy.

From the table, the total lumen output of the installation is 131,800 lumens.

The total circuit Watts of the installation is 2538 Watts.

Therefore, the average circuit efficacy is:

$$\frac{131,800}{2538} = \textbf{51.9 lumens/Watt}$$

The proposed lighting scheme therefore meets the requirements of the Regulations.

Note: If 100 W tungsten lamps were to be used over tables instead of the 60 W lamps actually proposed, the average circuit efficacy would drop to 43.4 lumens/W, which is unsatisfactory. If, however, 11 W compact fluorescent lamps, which have the same light output as 60 W tungsten lamps, were used over tables the average circuit efficacy would be 83.2 lumens/W.

Position	Number	Description	Circuit Watts (W) per lamp	Lumen output (lm) per lamp	Total circuit Watts (W)	Total lumen output (lm)
Over tables	20	60 W tungsten	60	710	1200	14,200
Concealed perimeter and bar lighting	24	32 W T8 fluorescent with high frequency control	36	3300	864	79,200
Toilets and circulation	6	18 W compact fluorescent with mains frequency control	23	1200	138	7,200
Kitchens	6	50 W, 1500 T8 fluorescent with high frequency control	56	5200	336	31,200
				Totals	**2538**	**131,800**

L

Standards referred to

BS 699: 1984 (1990) with amendments prior to June 1994 *Specification for copper direct cylinders for domestic purposes.*

BS 853: 1990 *Specification for calorifiers and storage vessels for central heating and hot water supply.*

BS 1566: 1984 (1990) *Copper indirect cylinders for domestic purposes.*

BS 3198: 1981 with amendments prior to June 1994 *Specification for copper hot water combination units for domestic purposes.*

BS 5422: 1990 *Methods for specifying thermal insulation materials on pipes, ductwork and equipment in the temperature range of -40°C to +700°C.*

BS 5449: 1990 *Specification for forced circulation hot water central heating systems for domestic premises.*

BS 5864: 1989 *Specification for installation in domestic premises of gas-fired ducted air-heaters of rated output not exceeding 60 kW.*

BS 6880: 1988 *Code of practice for low temperature hot water heating systems of output greater than 45 kW.*

Other publications referred to

Building Research Establishment (BRE)

BRE Report BR 262, 1994 *Thermal insulation: avoiding risks.*

BRE Report BR 265, 1994 *Minimising air infiltration in office buildings.*

BRE Information Paper IP 3/90 *The U-value of ground floors: application to building regulations.*

BRE Information Paper IP 7/93 *The U-value of solid ground floors with edge insulation.*

BRE Information Paper IP 12/94 *Assessing condensation risk and heat loss at thermal bridges around openings.*

BRE Information Paper IP 14/94 *U-values for basements.*

Chartered Institution of Building Services Engineers (CIBSE)

CIBSE Building Energy Code: 1981: Part 2 *Calculation of energy demands and targets for the design of new buildings and services: (a) Heated and naturally ventilated buildings.*

CIBSE Guide A: Design Data – Section A3: 1980 *Thermal properties of building structures.*

CIBSE *Code for interior lighting* 1994.

CIBSE Applications Manual AM1:1985 *Automatic controls and their implications for systems design.*

National House Building Council NHBC/Energy Efficiency Office

Thermal insulation and ventilation Good Practice Guide 1991.

L